THE LEAKEY FAMILY

Other Books by Delta Willis

THE HOMINID GANG: Behind the Scenes in the Search for Human Origins

FODOR'S TRAVEL GUIDE TO KENYA, TANZANIA & SEYCHELLES

Makers of Modern Science

THE LEAKEY FAMILY

Leaders in the Search for Human Origins

Delta Willis

New York • Oxford

THE LEAKEY FAMILY: Leaders in the Search for Human Origins

Facts On File, Inc.
460 Park Avenue South
New York NY 10016

Facts On File Limited
c/o Roundhouse Publishing Ltd.
P.O. Box 140
Oxford OX2 7SF
United Kingdom

Library of Congress Cataloging-in-Publication Data
Willis, Delta.
The Leakey Family : leaders in the search for human origins/by Delta Willis.
p. cm. — (Makers of modern science series)
Includes bibliographical references and index.
ISBN 0-8160-2605-X
1. Leakey, L.S.B. (Louis Seymour Bazett), 1903–1972–Family —Juvenile literature. 2. Leakey, Richard E.—Juvenile literature. 3. Anthropologists—Africa, East—Biography—Juvenile literature. 4. Fossil man—Africa, East—Juvenile literature. 5. Human evolution—Juvenile literature. I. Title. II. Series.
GN21.L37W55 1992
306′.092′2—dc20
[B] 92-12522

Facts On File books are available at special discounts when purchased in bulk quantities for businesses, associations, institutions or sales promotions. Please contact our Special Sales Department in New York at 212/683-2244 (dial 800/322- 8755 except in NY, AK or HI).

Text design by Ron Monteleone
Jacket design by Cathy Hyman
Composition by Facts On File, Inc./Grace M. Ferrara
Manufactured by R.R. Donnelley & Sons
Printed in the United States of America

10 9 8 7 6 5 4 3 2 1

This book is printed on acid-free paper.

CONTENTS

To my nieces and nephew, Jennifer, Mikki and Robert Willis

ACKNOWLEDGMENTS

I am especially grateful to Mary Leakey and Richard and Meave Leakey for their cooperation and permission to use photographs. Mary G. Smith and Neva Folk at the National Geographic Society were most helpful in providing permissions and prints for photographs previously published in *National Geographic.*

In addition to the references cited in the Further Reading section, I relied on extensive interviews and research conducted over nine years for various magazine articles and my book *The Hominid Gang*, published by Penguin. Part of that research included accompanying Richard and Meave Leakey, as well as other scientists, as they worked at West Turkana sites during the 1986 field season.

Thanks to Richard Milner, author of Facts On File's *Encyclopedia of Evolution*, for recommending me for this work.

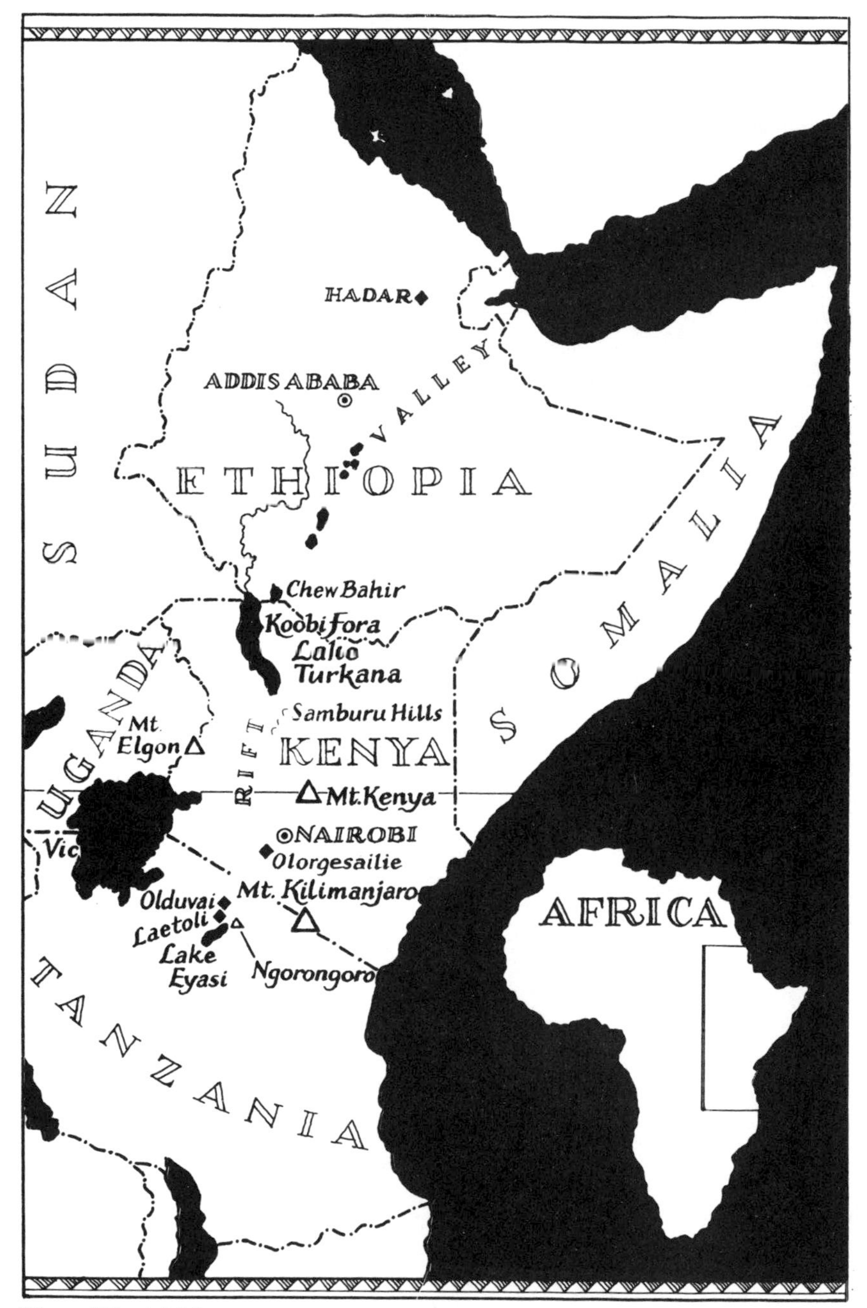

Map of East Africa (Bob Gale, © Delta Willis)

1

AFRICAN ROOTS

Richard Leakey was only six years old when he made his first important fossil discovery. The year was 1950, and the place was Kanjera, which is located in Kenya on the northeastern shores of Africa's largest lake, Victoria.

Richard had been told not to bother his parents, Louis and Mary Leakey, who were carefully excavating some fossils they had found. Yet he couldn't go play in the surrounding bush, because there might be lions or poisonous snakes in the area. So he quickly became impatient, hot and bored. Flies bothered him, and there was no tree nearby to provide shade from the hot African sun. When Richard began to complain—"I am thirsty; I have a stomach-ache; I am hot; what can I do?"—his father simply told him, "Go and find your own bone!"

Taking up this challenge, Richard began to wander around the excavation site. He had gone only about 30 feet away from his parents when he saw a small brown-colored bone, which he recognized as a fossil. He settled down on his knees and began to work.

Richard had his own tools: a dental pick, which he used to scrape away the bits of rock and dirt on the bone, and a small brush to gently sweep the debris away to the side. Fossil bones are not white like "new" bones of animals that have died recently, but absorb the color of the surrounding earth. As the brown bone began to emerge, he forgot about the flies and the heat and became totally absorbed by what he might be uncovering.

It was like digging for buried treasure, except that fossil discoveries are always something of a surprise. Sometimes they turn out to be the skeleton of an animal that no human being has ever seen before—a giant antelope with horns spanning six feet across, a giraffe with strange antlers like those of a moose or even a saber-toothed cat that lived in Africa 12 million years ago, before humans were here to see it.

The longer Richard worked the larger the brown bone appeared to be. Suddenly he saw one tooth, and then another. What was this animal that had been laying there for thousands and thousands of years?

Richard's silent industry finally attracted his parents' attention. When his father went over to investigate, Dr. Leakey saw that his young son had uncovered a jaw of a giant pig. The find was so important that both the boy's parents began to help with the excavation. The pig turned out to be much larger than the warthogs and forest pigs that roam Africa today; in fact, it was nearly twice the size! This particular species of pig became extinct and is only known by its bones. Richard's lucky find was such a good example

A family portrait—Louis and Mary Leakey during the 1950s, with sons Richard, Philip and Jonathan (left to right) and Mary's Dalmatians.
(Leakey archives)

that it earned an exhibit space in the National Museum of Kenya in Nairobi, where visitors still can see it today.

THE RECORD OF EVIDENCE

The prehistory hall of the museum is filled with other discoveries by the Leakeys, who are the most famous fossil finders in the world. Their discoveries include the most complete ancient skeleton ever found of our ancestors. It is known as the Turkana Boy, because it was found near Kenya's Lake Turkana; by studying the bones and the teeth, the Leakeys could tell that it was a male about 14 years old. The Leakeys were also the first to discover the fossil skulls and leg bones that belonged to the early toolmakers nicknamed "Handy Man." They found footprints that are the oldest record of our upright way of walking—which is what makes us different from all other primates.

The ability of humans to create art also distinguishes us from other primates, and the Leakeys studied Stone Age art in Tanzania that was created 1,500 years ago. Reproductions of these ancient drawings feature in the museum as well. (The original art is still on granite rock faces in Tanzania; Mary Leakey traced the drawings so that people visiting the museum might see what they are like.) While some primates, including chimpanzees, can make and use simple tools, the stone tools made by humans have much greater variety and more intricate patterns. The museum features many stone tools, including hand axes and small choppers. One of the largest exhibits is a display of a whole elephant, with its fossilized bones laying just as they were found. This cousin of our modern-day elephant belongs to the family of deinotheres (meaning "huge beasts") that lived over 2 million years ago, but its tusks curved down instead of up. And there is Richard's pig.

While the Leakeys are most interested in finding evidence of early humans, knowing about all the other animals and plants that lived on the same landscape in ancient Africa is vita. Some of the animal bones found alongside stone tools tell us what our ancestors ate. Distinct cutmarks made by stone tools appear on several antelope bones, for example.

Finding fossils isn't always as easy as it was for Richard that day at Kanjera. One reason why is because not all bones turn into fossils. First the bones must be buried naturally, by the sediments of a river or the ash of a volcano. Then the tiny holes of the bone must absorb surrounding minerals, such as calcium or silica. The chances that this might happen are rare, but in Kanjera and some other places in East Africa, there are so many fossils that you have to be careful where you step.

Because bones are first buried and then emerge slowly as heavy rains erode the soil, only a small fragment or edge might appear on the surface. Sometimes the rains are so heavy that fossils become completely exposed, and no digging is required. As Richard Leakey said recently, "Nature does most of the work." He was being modest. The fossilized bones of our ancestors are rarer than diamonds. The Leakey family and the teams that work with them have found the remains of hundreds of different individuals, and over two dozen skulls. The Leakeys have unearthed so many ancient finds of our early ancestors that East Africa became known as the "cradle of mankind."

The Leakeys are involved with paleontology; a science that has to do with prehistoric bones. (*Paleo* means "ancient.") But to be specific, they study human bones and stone tools to form ideas about how early humans lived and behaved. This science is known as anthropology. Because their study combines paleontology and anthropology, their field is called paleoanthropology. *Science Digest* magazine referred to the Leakeys as "paleoanthropology's first family."

Some people refer to their success as "Leakey's Luck." Luck comes to those prepared to meet a challenge, and this includes being in the right place at the right time.

THE AFRICAN HORIZON

The Leakeys live and work in Africa. Louis and Richard Leakey were born in Kenya, and Mary and Meave Leakey spent most of their adult lives in Africa. Louis Leakey died in 1972, but Mary, Richard and his wife, Meave Leakey, continued to live and work in Africa as of 1992, when this book was written. Living there may seem like a great adventure that has little to do with their scientific work. But it gives them an edge over other scientists who live in modern cities in the United States and Europe, and who visit Africa only occasionally to do their field research.

Living in Africa provides scientists with a very keen sense of natural history. The Leakeys often work in remote, desert landscapes, where they can observe a world that is not so different from how it might have been thousands of years ago. For example, when Mary Leakey worked at Olduvai Gorge in Tanzania, she saw the great migration of millions of wildebeeste that move across the Serengeti plains every year. This, plus some careful scientific research, inspired a theory of how our ancestors might have moved with these herds, following them as a handy food supply.

Also, her camp at Olduvai was surrounded by nomadic Maasai, a people who at that time were still living a traditional way of life. Tanzania is also home to the Hadza people, whose women gather foods in an old-fashioned way, digging up tubers with a stick, or walking long distances to gather certain berries. It's quite different from the life the rest of us know. While we drive to the grocery store for our potatoes or fresh berries, the Hadza have to know when certain plants are in season and where to find them, and must organize their gathering, by foot, accordingly.

While the people of Africa today are certainly not the same as our ancestors, they can give clues and ideas about what it is like to rely on the landscape for food, water and shelter. Most of the nomadic people in Africa construct their homes out of thorn trees, mud and cattle dung. They also have great knowledge about which plants are of medicinal benefit. The Leakeys speak the languages of their neighbors, and by knowing about the traditional ways of life that still exist today, they might have gained some important insights into our past.

Observing other primates, such as our cousins, the chimpanzees, also provides clues to our ancestors. It was Jane Goodall's careful study that proved that chimps use tools, but it was Louis Leakey's hunch that something could be learned from these primates that inspired her in the first place. In Africa, primates—especially baboons and vervet monkeys—can be seen around camp on a daily basis.

Africa also provides daily reminders of geological changes—such as the many volcanoes of the Great Rift Valley—as well as different weather patterns. Both influenced the shape of the lakes and the flows of the rivers where our ancestors lived and where most of the fossils and artifacts are found. You can see a lot of this landscape just by driving through the area in a car, but you can see the big picture from a small airplane. Richard Leakey flies one all over Kenya, and he has found many fossil sites by looking out the window.

When you are in the African bush, it is not unusual to see a predator stalking its prey or the remains of a lion kill. If you are in one area for a long time, you can watch what happens to the bones, how hyenas come along and scatter them. This gives you a good idea of why very few skeletons of wild animals are found whole, or intact.

Even when you find a fossil skeleton in a certain place, that is not necessarily the same place where the animal died. In 1968, when Richard Leakey was working in Ethiopia's Omo River Valley, north of the Kenyan border, he noticed a dead warthog floating downstream. He wondered where the pig would finally land and where its bones would be buried by the river sediments. This served as a good reminder that fossilized bones don't always mark a place where an animal actually lived or died.

All of these observations help to answer the questions that arise when discoveries are made. Of course, visitors to Africa can see some of these things, and many scientists do accomplish a great deal on their short visits. But when you live in Africa, everyday you can observe the clues around you. Living in cities, we are distant from nature. The Leakeys prefer to live in the African bush.

A CLASSROOM WITHOUT WALLS

The Leakeys' research camps are very simple but efficient. Green canvas tents or simple stone huts with thatched roofs can be home for a long time. There are no piped water supplies in the bush, so often one bathes in a river or a lake (keeping an eye out for hippos and crocodiles). Nor is there any electricity, which means no television, no videos and no computer games! Nor are there telephones. (Battery-operated computers are sometimes used for research, but sparingly.)

At first, this setup may seem like a disadvantage, but once you have decided to take up a challenge, the next major step toward success is concentration and focus. This means dedication to that task and little else. The camp schedule is busy and long, with everyone rising before the sun is up. Workers search for fossils early in the morning, when it is cooler, take a break at midday, and then resume the search again in midafternoon, after the worst heat of the day.

During the evenings, when other forms of entertainment could prove distracting, the Leakeys talk to each other and members of their research team—about fossils. They discuss what has been found and what it might mean. One of the members of the team might bring a fossil to the dinner table as a clue and say, "What do you think this is?" The dinner table at camp is like a wonderful classroom without walls.

Living in Africa also invites a world of adventures. While the Leakeys do spend hours in a museum laboratory in Nairobi studying their finds, and they do have "normal" homes in the suburbs of Nairobi, some days in the field are like an Indiana Jones movie. For a while, the daily adventure included riding camels while searching for fossils.

It was in 1969, on one of these camel expeditions, that Richard and Meave Leakey made a very unusual discovery. They were camped out near Koobi Fora, which is a sandy peninsula on the eastern shore of Kenya's Lake Turkana. They had set out early in the morning, and by late morning Richard decided to return to camp by taking a shortcut. They walked along a dry riverbed, leading their camels, but came to a sudden halt.

Richard saw a skull, sitting on the slopes of the riverbank. Finding a complete skull like this is very rare; most of the time, discoveries are found in many different little pieces, and the skull has to be put together like a jigsaw puzzle. But here was a complete skull!

Richard gazed at this skull as if he had recognized an old friend's face. It was just like a skull that his mother, Mary, had found in

GEOLOGICAL TIME SCALE

MILLIONS OF YEARS BEFORE THE PRESENT	PERIOD	EPOCHS	HOMINIDS
		CENOZOIC ERA 65 million years	
0.01	QUATERNARY	HOLOCENE	Neandertal 70,000 yrs. ago; H. erectus 1.6 M.Y.A.
2		PLEISTOCENE 2 MILLION YRS. AGO	H. habilis; A. africanus; A. robustus A. boisei
5	TERTIARY	PLIOCENE	A. afarensis 3.-3.5 m.y.a MILLION YRS. AGO
24		MIOCENE	
38		OLIGOCENE	
55		EOCENE	
		PALAEOCENE	
65	PERIOD	MESOZOIC ERA 175 million years	
140	CRETACEOUS		
205	JURASSIC		
	TRIASSIC		
240	PERIOD	PALEOZOIC ERA 330 million years	
290	PERMIAN		
330	CARBONIFEROUS	PENNSYLVANIAN	
360		MISSISSIPPIAN	
410	DEVONIAN		
435	SILURIAN		
500	ORDOVICIAN		
	CAMBRIAN		
570	PRE CAMBRIAN TIME About 400 million years		

Geological time scale (Bob Gale, © Delta Willis)

Olduvai Gorge, in Tanzania, exactly 10 years earlier. Richard recognized the details—the large teeth, the big face, the small braincase—as one of our ancient cousins. It has been given the name *Australopithecus*, which means "southern ape." Is it one of our ancestors?

THE SEARCH FOR MISSING LINKS

Humans have always wondered about the origins of our life on earth. When and how did we get here? Where were humans when dinosaurs roamed the earth?

If we want to discover what the earth was like before there were written records, then we must turn to the record that lies buried in the earth. This includes the fossil record and the geological record, which helps us date the fossils. From this, we know that about 150 million years ago, the earth was home to many forms of dinosaurs. But with all the thousands of dinosaur bones that have been found, not one humanlike bone has emerged from this time frame. And there has been a log of digging, all over the world. So we know that human beings did not exist at the same time.

This may be easy for us to visualize now, because so much information on the dinosaur era is available. We can see the real bones of these dinosaurs in museums, and films and videos bring their story alive for us and allow us to imagine a different world.

But only 150 years ago, human beings had quite different ideas about life on earth, because they didn't have this information. As we look back at these ideas, it is important not to laugh at people for thinking the way they did—even though some of their ideas may seem silly to us. After you laugh anyway, remember that we are making similar mistakes right now, and 150 years in the future, students may have a laugh at our ideas! Testing ideas is the challenge of science, and Louis Leakey was bold about testing ideas. His college advisers told him not to bother looking in Africa for clues to human ancestors, but Leakey followed the hunches of Charles Darwin, whose theory of natural selection was vital to explaining how humans evolved from ape-like ancestors. Paleoanthropology is a very young science when compared to other disciplines, such as astronomy, and many ideas remain to be tested. If you decide now to become a fossil finder, you will still have plenty of research to do throughout your lifetime.

You may already know that we are part of the animal kingdom and that scientists believe we evolved from much smaller pri-

Richard Leakey with the complete skull discovered at East Turkana in 1969. This *Australopithecus* is the same species as the "Zinj" his mother Mary had found 10 years earlier at Olduvai Gorge in Tanzania. (Meave Leakey, © *National Geographic*)

mates that lived in the forests of Africa. When you go to the zoo and see chimpanzees, orangutans and gorillas, you can see why we are all in the same order of primates.

Many people think that humans are present on earth because of supernatural forces, that we were put here by God. Others think that humans arrived on earth by extraterrestrial means, that our forebears sprang from another planet, or at least another place in the universe. These ideas are not totally invalid; they simply have yet to be proved.

By exploring the earth's fossil records, we do have proof of evolutionary change among hominids. The *Australopithecus* skull that Richard and Meave Leakey found in 1969 is just one example of this evidence. The *Australopithecus* is a form of primate that marks an in-between stage of apes and humans. It has many apelike characteristics (such as a small brain) and many human characteristics (it walked upright on two legs). It is 1.8 million years old. When this species was first discovered, its age came as a great surprise. This discovery was one in a series by the Leakeys that pushed our origins far back into the past. Understanding how much time evolution requires is important to understanding the whole process of our origins.

Louis Leakey recognized the importance of time, and in addition to studying geology himself, he was the first paleoanthropologist to include a geologist as part of the team in the field, as Richard, Mary and Meave do today.

In 1830, geologist Charles Lyell published the first volume of his monumental *Principles of Geology*. This work proved that the earth was much older than anyone thought and that there had been time for tremendous changes among plant and animal life.

Shortly after Lyell published his books, a young Charles Darwin set out on the voyage of the *Beagle*, to explore plant and animal life in the Galápagos Islands and in South America. He took along Lyell's books, which helped explain what he saw. The idea of evolution—changes among species through time—was not new. But what Darwin eventually did develop was an idea for how evolution worked, called natural selection. When he explained how certain features changed, he wrote that "Light will be cast on man and his origins."

When Darwin published this in *Origin of Species* in 1860, most people thought it was outrageous. They believed that humans were separate from nature and certainly not related to apes, as he had suggested.

A few years later, in 1871, Darwin wrote in *The Descent of Man* that Africa was the place to look for fossilized "missing links," in-between forms of humans and apes. Darwin reckoned that missing links might be found there because that is where our cousins—the chimps and the gorillas—lived. But many scientists preferred to look anywhere but Africa for our origins, and searched in England and Germany, in Java and in China.

Louis Leakey was born in Africa in 1903. At a very early age, he decided to look for evidence of our ancestors in his own backyard.

CHAPTER 1 NOTES

p. 1 "Go and find your own bone . . ." Richard Leakey, *One Life*, p. 28.

p. 4 "Paleoanthropology's first family," "Discovering Africa's Ancient Art," *Science Digest*, August 1984, pp. 56–63.

p. 11 "Light will be cast . . ." Charles Darwin, *On the Origin of Species by Means of Natural Selection*, p. 222.

pp. 12–13 "missing links . . ." Charles Darwin, *The Descent of Man*, p. 138.

2

WHITE AFRICAN

Louis Leakey is often described as a "British" anthropologist. Yet Louis took great pride in his African roots and, from a very early age, was determined to prove that human life began in Africa.

Louis Seymour Bazett Leakey was born in Kenya to missionary parents. His father, Harry, and his mother, Mary Bazett, had gone from England to work for the Anglican church. Louis' first home was a simple house made of mud and thatch, with a dirt floor. The Leakeys lived in a small town called Kabete, eight miles north of Nairobi.

Today, Nairobi is a modern city, with many spectacular hotels and four-lane highways. When Leakey was born in August of 1903, it was only a small stop on the East African Railroad line; the streets were dirt. A cemetery had been established for victims of man-eating lions, and the local doctor made his rounds on a saddled zebra. The 50 ethnic tribes of Kenya still lived a traditional way of life, and Louis was to be greatly influenced by local Kikuyu, one of the larger tribes, engaged in agriculture in the highlands.

Shortly after his birth, members of the local Kikuyu tribe walked for miles to see him. Many had never seen a white baby before, but they showered him with one of their traditions, spitting! His mother was horrified, and fearing potential infection, Louis was immediately washed clean. Yet the Kikuyu meant it as a show of trust. Many Africans believe in witchcraft, and to cast a spell, the witch has to have a piece of the victim's body, such as hair. To show no evil intent, many Africans offer their saliva. It was Louis' first initiation into the ways of the Kikuyu. He soon became a convert of the very people that his parents were trying to convert.

MEMBER OF THE KIKUYU TRIBE

As a young boy, Louis followed his Kikuyu playmates into the bush, learning how to survive in the wilderness and developing a lifelong interest in wildlife. He spoke Kikuyu as fluently as English and spent many evenings listening to the legends that the elders told. At the age of 13, he was initiated into the Kikuyu tribe as a respected member. The chief of his tribe, Koinange, referred to Louis as "the black man with a white face."

Louis' identification with the Kikuyu would later shape his professional career. As an adult, he would write three volumes on the anthropology and customs of the Kikuyu and a grammar book on their language. His knowledge of their language eventually involved him in the politics of the "Mau-Mau" rebellion, which led to Kenya's independence from England in 1963.

Such was the influence and bond with his Kikuyu friends that, throughout his life, he thought in Kikuyu rather than English; in fact, Louis wrote that he dreamed in Kikuyu. He entitled an early autobiography *White African*.

Louis also learned to speak Swahili, which today is Kenya's national language, and stalked antelope with a member of the Dorobo tribe, Joshua Muhia. "From Joshua I learned to camouflage my human form with leaves and small branches; to approach a quarry diagonally, and very alertly, and above all, never to show my hands or arms," Louis recalled in a book for the National Geographic Society. "Most important, I learned patience, for it was necessary to get quite close to an animal if it was to be killed with the Dorobo's short-range weapons—the bow and arrow, thrusting spear, or club." As a paleoanthropologist, "I have used much of what I learned when trying to interpret the possible ways Stone Age man hunted and trapped his prey." (Louis was not a "white hunter" shooting game for trophy; in fact, he became one of Kenya's leading conservationists, helping to establish the national parks.)

"As a result of his early training and interest," Richard Leakey wrote of his father, "he became a scientist rather than a missionary." It is not so ironic that Louis Leakey would grow up to follow Charles Darwin's advice—to look for missing links in Africa—even

Louis Leakey at work at Olduvai Gorge, holding a fossilized elephant tooth in his right hand and the broken molar of an extinct dinotherium in his hat. (Melville Bell Grosvenor, © *National Geographic*)

through he was surrounded by a family that might find Darwin's ideas shocking. Darwin himself grew up in a religious family. In fact, he didn't publish his theory for over 20 years, as he was afraid of offending his wife and many other religious people in Victorian

England. Such was the outrage following the publication of *Origin of Species* in 1860 that the wife of the Bishop of Worcester is said to have exclaimed, "My dear, descended from the apes! Let us hope it is not true, but if it is, let us pray that it will not become generally known."

Louis Leakey, as a public speaker and prolific author, would do a great deal to make our kinship with apes known to the general public. Dr. Leonard Carmichael of the National Geographic Society described Leakey as "the Darwin of human prehistory."

Young Leakey, like young Darwin, was inquisitive about the world around him. Inevitably their observations led to new discoveries. Much of their genius involved looking at the same old things in new ways.

DISCOVERING STONE TOOLS

In 1915, when he was 12 years old, Louis received a book entitled *Days Before History* as a Christmas gift from a cousin in England. Written by N. H. Hall, it described the evidence for the Later Stone Age in Europe, where most of the stone tools had been found thus far. The Stone Age was also known as the Flint Age, because most of the tools found in Europe were made of a stone called flint—a variety of chert that produces sparks when struck, and was therefore used to start fires as well as shape tools. Louis found the book fascinating, and immediately set out to search for arrow points and ax heads near Kabete. Louis didn't know there wasn't any flint in the area; there is no flint in all of East Africa. But he was inspired, and searching along the dirt roads and areas where the earth was exposed, he found plenty of tools.

They were made of a volcanic rock called obsidian. Obsidian is a glassy-looking black stone; flint is also black, but dull. The shapes and the patterns that Louis found were so similar to the ones in his book that he collected these tools and kept them "with religious care." His parents were skeptical; whatever these things were, they told him, they weren't flint tools.

His Kikuyu friends laughed at him for suggesting that these flaked stones were the work of people who lived long ago; they had their own explanation: They were *nyenji cia ngoma*—the discarded razors of spirits of the sky. Louis found this quite understandable, if incorrect: "Most of the prehistoric obsidian implements and flakes in the Kikuyu country are buried fairly deeply in the soil . . . After a very heavy fall of rain, pieces of obsidian are particularly noticeable, so it is not surprising that the Kikuyu think that they have come down out of the sky with the rain."

Eventually he showed his collection to an expert at the Nairobi Museum, Arthur Loveridge, who was his hero and mentor. The museum curator confirmed that they were indeed tools and encouraged Louis to keep a careful record of his finds. So, at the age of 13, Louis "embarked upon a study of the Stone Age in East Africa," determined to learn all he could.

Only four years earlier, in 1911, two American archaeologists had visited Kenya in search of Stone Age tools. But because they were used to the flint examples, they didn't see any—or so they thought. Amazingly, they searched around Lake Naivasha, one of the lakes in the Rift Valley, west of Nairobi, an area that is rich in obsidian tools.

"I believe they actually stayed at a Naivasha hotel," Louis Leakey wrote in *White African*, "where the whole surface of the ground is covered with obsidian flakes and tools. . . ." But they went home empty-handed, convinced that Kenya held no real potential for archaeological exploration. Perhaps not knowing what flint was was an advantage, for young Leakey recognized the tools by looking for shapes and patterns, which was key. (Descriptions of the distinctive patterns are in Chapters 3 and 4.)

STUDIES IN ENGLAND

Because there was no elementary school nearby, Louis learned math, French and Latin from his father and tutors who were brought to Kabete. At the age of 16, he was sent to England to study. His learning experiences in Africa scarcely prepared him for the demands of a formal classroom; when one of his instructors assigned an essay, Louis went up to him after class and asked what an essay was. He was also rather lonely, and developed few friendships like those he had known among his friends in Kenya. But he was determined to succeed and to study anthropology at Cambridge University.

One of the college requirements was knowledge of two foreign languages. Leakey had learned French, and for his second language he suggested Kikuyu. At first the officials balked at this extraordinary proposal, but there was no rule against it, and Leakey argued his case forcefully. Because no instructor at Cambridge knew Kikuyu, Leakey taught the language to his supervisor, who in turn tested his young instructor. It was a clever solution to a sticky problem, and Louis was very clever in finding ways to get what he wanted.

At Cambridge, one of his anthropology professors, Dr. A. C. Haddon, made himself available to students at his home on Sunday afternoon. Louis seized this opportunity, while many other students did not. This is yet another example of the many ways in which Louis moved ahead, learning as much during these informal sessions as during the lectures, and borrowing anthropology books from Dr. Haddon's library. Haddon led Louis to become interested in the string figures known as cat's cradles. People created figures and told stories by dipping their fingers into a maze of string to create a pattern. Louis later investigated cat's cradles among African tribes and published a book on the subject.

His studies at Cambridge were interrupted in his second year, when he suffered a concussion while playing rugby. Afterward, he had severe headaches when trying to read and briefly lost his memory; his doctor recommended that he take a year's leave of absence. Louis filled the time in a creative and fruitful way, by helping to organize a scientific expedition to Tanzania (then the German colony of Tanganyika). This job assignment resulted form what is known today as "networking."

Leakey used every potential contact; some came from family connections, some were merely names in newspaper articles. He was not shy about introducing himself or calling upon scientists who were several years his senior.

Leakey's knowledge of Africa and his ability to speak the languages were useful to the expedition of the British Museum of Natural History, led by W. E. Cutler. Cutler had never been to Africa, but he planned to collect dinosaur fossils that had been found at Tendaguru, west of the southern port of Lindi, Tanzania.

Cutler's expedition was a wonderful opportunity for Louis, who gained experience excavating and preserving fossils.

(Sometimes fossils are fragile and must be painted and preserved with a strong glue. They must also be packed carefully for transport to museum archives.) Louis celebrated his 21st birthday in Tanzania and, on the return journey, trekked 269 miles to Dar Es Salaam to catch a boat home to England.

When he returned to Cambridge, he began to study geology in addition to his other courses, because he had found a knowledge of geology so useful on the Tanzanian expedition. It was the beginning of his multi-disciplinary approach, and he would eventually employ this approach in the field, combining the disciplines of paleontology, anthropology, primatology, geology, anatomy and archaeology.

He also found creative ways to solve his continuing financial problems. His parents didn't make very much money in Kenya, and Louis relied on grants, scholarships and small projects to help pay his college fees. Once he worked as a housekeeper and cook, and another time he brought back to England 100 ebony walking sticks made by Tanzanian carvers, which he bartered with tailors in Cambridge. He also received payment for writing articles and giving lectures about his expedition with Cutler.

His first lecture showed his genius for captivating an audience. While it is very common now to have slides or a film clip with a lecture, it wasn't at that time. Not only did Louis have these visual aids, but he had obtained models of dinosaurs from the producers of a movie called *The Lost World.* Louis assured the producers that the publicity would be good for their movie, and he borrowed the props for free.

He was so nervous about the lecture that he spent much of the day in a cold sweat, but when the lights dimmed and he began to speak, all went well—for a while. Then something distracted him, he remembered his fears, and the rest of his speech was halting. After this first lecture, however, he became more confident and grew to love public speaking. He was able to inspire all sorts of audiences, from young children to distracted executives.

At Cambridge, he wrote his thesis on the Stone Age evidence in Kenya and was graduated with highest honors. "The dreams that I had dreamed as a child after reading Hall's *Days Before History* were coming true," he wrote in *White African.*

With his good performance in college and his experience in the field, Louis gained grants and other financial support for several expeditions to East Africa. His explorations included gathering stone tools at a place in Kenya known as Kariandusi, but the most important destination was a dramatic gorge called Olduvai.

THE FIRST TIME TO OLDUVAI

Olduvai Gorge is located in northern Tanzania, between the majestic volcanic caldera known as Ngorongoro (*n-go-row n-go-row*) Crater and the Serengeti National Park. Like the Grand Canyon, Olduvai was cut by a river. Many tools and fossils were exposed in the gorge, which is 300 feet deep and 25 miles long. Part of the floor of the gorge once held a lake.

The geology of Africa is complex, especially in the area known as the Great Rift Valley. Dramatic tectonic (earth-moving) and volcanic activity caused huge changes in the earth's surface, and the tremendous amount of alkaline substances that volcanoes emit also make East Africa and Ethiopia ideal for finding fossils. The alkaline helps to preserve bones that might otherwise simply disintegrate, and the shifts in the earth's surface tend to expose the fossils many years later. These factors combine with the influence of flooding rivers, a result of the seasonal monsoon rains.

The rich fossils of Olduvai Gorge were first noticed in 1911 by a German Professor Kattwinkel; legend has it that Kattwinkel was chasing a butterfly that led him to the brink of the gorge and nearly fell over the edge. (Louis loved to tell this story.) Kattwinkel slowly climbed down to the floor of the gorge where he discovered fossils of a three-toed horse. This inspired a German expedition to the site led by Hans Reck, who found 1,700 fossils but no tools—or so he thought.

The way that Louis made his way to Olduvai refines the idea of "Leakey's Luck." He was prepared for a challenge; he made connections with the leading people in his field; he was in the right place at the right time.

Ironically, many other scientists were certain that human origins began in Asia. A Cambridge professor had discouraged Leakey from searching for missing links in Africa: "Don't waste your time. There's nothing of significance there. If you want to spend your life studying early man, do it in Asia." Leakey's reply was confident and determined: "No, I was born in East Africa, and I've already found traces of early man there. Furthermore, I'm convinced that Africa, not Asia, is the cradle of mankind." Much later in his life, he would recall that he "decided at the age of 13 to find out if Darwin was right."

In 1927, Louis decided to do some research that required a trip to Germany; he obtained a grant and sold his motorcycle to help fund the trip. In Berlin, he looked up Hans Reck, and the two talked about the fossils of Olduvai. Louis noticed that one of Reck's "rock samples" from the gorge resembled some stone tools that he had found on his own expedition to Kariandusi, in Kenya. Leakey invited Reck to join his 1931 expedition to Olduvai, betting him £10 that he could find a tool there within 24 hours. Leakey won the bet, because Reck (like the American scientists at Naivasha) had been searching for flint tools.

The tools from Olduvai were very similar to those that Leakey had found at Kariandusi. Because of this similarity, Louis agreed with Reck regarding the age of a human skull and skeleton that Reck had found on his first Olduvai expedition. Another fossil finder, A. T. Hopwood, who had found and named an early primate from Kenya, also agreed. Within a week of their arrival at Olduvai, in September of 1931, the three men sent a letter to a scientific publication in England, saying that they had solved the mystery of the so-called Olduvai Man, which they claimed was much older than it turned out to be. When independent geologists came to date the site, it turned out that Leakey and his team were wrong; the bones Reck had discovered had belonged to a modern human.

Mistakes, no matter how embarrassing, do happen. But when Leakey made a second mistake regarding the age of a fossil, it would lead to greater controversy, because it appeared that he was being careless.

CONTROVERSY

Following the Olduvai expedition, Leakey returned to western Kenya and a site called Kanam, which is near Homa Bay on Lake Victoria.

At Kanam he found a fossilized jaw with several teeth. Leakey thought the jaw represented the oldest evidence of our ancestors.

By this time several other major discoveries had been found in other parts of the world, so Leakey's claims were bold. In 1925, anatomist Raymond Dart had reported on a small skull that was discovered in South Africa. It was known as the Taung child, because it was discovered in the Taung quarry, where workmen were mining limestone. Dart suggested that this apish-looking little skull represented one of our ancestors, and called the find an australopithecine. (*Austral* means "southern," for southern Africa, and *pithecine* indicates the ape family. Humans and apes are both in the primate order.) Dart's claims happened to be correct, but they were ignored by most scientists, including Leakey, who never thought australopithecines were ancestral to humans.

In 1927, skulls and skeletons were found in China near Peking (Beijing), and they were called Peking Man. (The search for early *man* also included the search for early women, just as the category "Peking Man" included fossils of females.) Leakey dismissed these finds as those of a cousin that became extinct rather than an ancestor of the human lineage.

Yet the finds from China and South Africa did indeed belong on our family tree, and they were legitimate: Ironically, Louis questioned the legitimacy of Peking Man when he should have been concerned about questions regarding the legitimacy of his own finds.

British geologist Percy Boswell suggested that more evidence of the geology and other fossils from the Kanam area were needed. Not only was the jaw that Leakey was using as evidence incomplete, fossils of other animals found in the same level of earth often help scientists date a find.

Leakey responded by inviting Boswell to go to Kenya and see the site for himself. It might have been a good solution to the challenge, if only Leakey had rehearsed the test survey carefully.

When they arrived near the excavation site, Louis couldn't find the exact boundaries, which had been marked by metal pins. Local fishermen had removed the pins—they were ideal fishing hooks. It is always essential to photograph discoveries as they are found, but Leakey's camera had not worked correctly during the 1932 excavation, and his photographs were no good for reference. He borrowed someone else's photographs, and to his horror they turned out to be mislabeled. Leakey had just sent one of these photos off to be included in a publication, and he had to stop the presses in a hurry. It was a nightmare.

Boswell was a man in search of careful, detailed work, and he left with a bad impression of Leakey. The geologist returned to England to write a critical paper that was published in the influential journal *Nature*. Even though the value of the jaw itself was not disproved, Leakey did not appear to be a careful or meticulous field scientist. This Kanam episode, as well as the mistake on the age of the Olduvai skull, damaged Leakey's reputation.

In 1935, the same year that the Boswell paper was published, Louis left Cambridge yet again to return to his work at Olduvai Gorge. One member of his team was a young British artist named Mary Nicol, who had illustrated Leakey's book, *Adam's Ancestors*. Mary had good experience working in the field, particularly with finding stone tools, and Louis was very impressed with her drawings, and her. Louis later separated from his wife, Frida, shortly after she had their second child; they were divorced in 1936, and Louis promptly married Mary. Leakey's biographer, Sonia Cole, suggests in *Leakey's Luck* that the divorce, combined with the Kanam controversy, ruined Louis' chances of becoming a professor at Cambridge, and inspired Louis and Mary to leave England and live in Africa. Yet their move to Africa was the beginning of a great partnership in discovery, and Mary became Louis Leakey's greatest find.

CHAPTER 2 NOTES

p. 14 "The black man with a white face . . ." Melvin Payne, "The Leakey Tradition Lives On," *National Geographic*, January 1973, p. 143.

p. 14 "From Joshua I learned . . ." Louis Leakey, *Animals of East Africa*, pp. 9–12.

p. 14 "As a result of his early training . . ." Richard Leakey, *One Life*, p. 16.

p. 16 "My dear, descended from the apes!" Richard Leakey and Roger Lewin, *Origins*, p. 21.

p. 16 "the Darwin of human prehistory," "The Leakey Tradition Lives On," *National Geographic*, January 1973, p. 143.

p. 17 "Most of the prehistoric obsidian implements . . ." L. S. B. Leakey, *White African*, p. 188.

p. 17 "embarked upon a study of the Stone Age . . ." L. S. B. Leakey, *White African*, p. 70.

p. 17 "I believe they actually stayed . . ." L. S. B. Leakey, *White African*, p. 81.

p. 19 "The dreams that I had dreamed . . ." L. S. B. Leakey, *White African*, p. 160.

p. 21 "Don't waste your time . . ." Transcript of an interview with Keith Berwick, 1959.

p. 21 "decided at the age of 13 . . ." Louis Leakey, "Family in Search of Man," *National Geographic*, February 1965, p. 214.

3

ADAM'S ANCESTORS

That Mary would develop a keen interest in stone tools is part of her own legacy. Her great-great-grandfather had found flint tools in Suffolk, England. In fact, he was the first Englishman to recognize them for what they were.

In 1797, John Frere delivered a report to the Royal Society in London that suggested these tools were made long ago, because they were unearthed at a depth of 12 feet. Depth can indicate age, because things found on higher levels of earth are usually younger than those found on lower levels. At neat sites you can start at the top and trace the recent to the older by moving lower, in levels of earth that look like layers of a cake. This is especially true in England, Europe and North America, where there has been less upheaval and change than in the Great Rift Valley of Africa. There many sites have levels that became mixed up and out of order. Dates for stone tools are fixed in a number of ways, including studying fossils in the area.

Frere couldn't put an exact date on his finds, but he suggested that humans may have made these tools in "a very remote period indeed, even beyond that of the present world." But his suggestion was dismissed, for it came at a time when people still thought the age of the earth was very young. (This was before Charles Lyell published his books on geology in 1830, which proved that the earth was much older than 6,000 years.) Most people believed in the biblical story of Adam and Eve, which had it that humans arrived rather suddenly on earth, and in a modern form. Now there is plenty of fossil evidence to show the very different forms of Adam's (and Eve's) ancestors.

While several people had found stone tools in the 18th century, many Europeans thought they were the result of lightning bolts striking the earth. The same explanation was once used for fossils. It seems funny now, but we must remember that the fossils must have looked very strange; perhaps many were plants and animals that no one recognized because they were extinct forms, and were all the more strange looking because only parts of them were found. Tools might be noticed as different, but few people imagined that they might be human creations. These lightning-bolt explanations were simply an attempt to explain the unknown, just as were the Kikuyu's "razors from the sky."

Frere's insights were ignored, and his ideas were considered heretical. A century later, when everyone realized that Frere was right, his name became famous in British archaeology.

KNOWING TOOLS WHEN YOU SEE THEM

How can you tell a stone tool from a rock? At one of the best colleges in the United States for paleoanthropology, the University of California at Berkeley, archaeology students are shown rocks from the local parking lot mixed in with some stone tools. Selecting tools from trash is a difficult test, especially when the tools are primitive.

Rocks break naturally all the time. Many split as a result of changes in temperature; the sun heats them up during the day, and during the night, they cool. But such natural breaks usually create an accidental-looking shape, with no real design or pattern.

Early humans might have picked up and used naturally broken rocks, just as you might use a rock in some way yourself, to dig or to throw at a target. It is interesting to think about how humans got the idea for making tools in the first place and how they might have discovered that they could flake certain kinds of stone and lava, such as obsidian. The earliest hand axes appear in Africa in levels dated at 1.6 million years old; this is also the date for the oldest known *Homo erectus* ("upright human," walking on two legs) found in Africa. But Mary Leakey has found even older tools in Tanzania, over 2 million years old.

Mary Leakey at her Langata home in Kenya with two of her many Dalmatians. The photograph was taken in 1984, shortly after Dr. Leakey finished her fieldwork at Olduvai Gorge in Tanzania. (© Delta Willis)

It is difficult to try to pinpoint exactly when humans first used stone tools. They probably used sticks and branches as well, but these wooden tools are rarely preserved, and most have been lost over the centuries. Some bone and wooden tools have been found in Spain, but the site there is much younger than the African sites. Fossilized fishing nets have been found near Lake Turkana in Kenya. As more research is done, more will be known.

NAMES FOR TOOLS

Stone tools appear in many forms. Many are named after what appears to have been their function—choppers and chisels, hammerstones and anvils, cleavers and scrapers. On some small, thinner tools, you can see a deliberate series of little serrations, or sharp ripples, like the edge of a bread or steak knife. The classical tear shape of an Acheulian hand ax is very easy to recognize because of its design and pattern. (*Acheulian* is pronounced uh-'SHOO-lee-un, and named after St. Acheul in northern France, where these tools were first found in abundance.)

Stone tools are made by striking a piece of lava or flint with another hard object, usually a harder rock. This process is called flaking, because little flakes fall off, leaving a concave dent, or scar, with each strike. These concave scars occur in a deliberate pattern or design. Often when looking at a tool you can see a symmetry that could result only from flaking. By drawing stone tools, Mary Leakey gained insight as to how they were made. "I had been handling stone tools since I was 11," she said a few years ago. "They follow a given pattern."

Archaeologists in the field often do experiments where they actually make and use tools. Louis Leakey was an expert and a showman at this; he would look around for the right kind of stone, pick up a good piece and begin to flake away. He practiced this so many times that he could prepare a useful tool within four minutes. Louis liked to test ideas in hands-on experiments rather than just proposing ideas about what early humans did with their tools. So he sometimes skinned different animals with different tools to see how well they worked. He once even tried to skin a hare with his teeth, to see if early humans could have done this rather than using tools; he concluded that they couldn't.

Mary Leakey was more interested in finding tools and drawing and studying them than in these kinds of experiments. But one of her young assistants, Peter Jones, followed Louis' example and did tests skinning and butchering animals. When a tool became dull when he tried to skin a big animal, Peter just knocked off a few flakes from both sides, giving them sharper edges. This left many flakes on the ground.

Flakes were first defined as tools, but now it appears that some may just have been by-products. When a huge number of flakes were found at kill sites (along with the fossil bones of a dead animal), scientists began to think that perhaps early tool users were simply sharpening their tools as they worked, as Jones had done. Flakes, which can be as sharp as razors, are useful in the last stage of butchering. So perhaps one large piece of lava was used in two ways, which is clever.

If you live near a good museum, you can probably see stone tools on display. American artifacts are quite different from those found in Africa, however. Indian arrowheads and spears feature refined work, while some of the tools from Africa are crude. Of course American natives appeared much later on the scene than our early ancestors, by several million years. The first Americans may have migrated here across a land bridge at the Bering Strait near Alaska, perhaps as recently as 12,000 years ago.

These North Americans are thought to have been in the last wave of a migration that began in Africa, moved into parts of Eurasia and eventually filtered on down to South America. Another major path in the migration from Eurasia was through Indonesia and onto the northern territory of Australia. The story is complex and involved much time and many paths, but these migratory routes offer a fair suggestion of the way humans began to dominate the earth, based on the human fossils and tools that have been found. When new discoveries are made, the story may alter in details and dates, but that is the nature of anthropology. Ideas are still being refined.

The patterns and designs of stone tools are well known today, based on the study of thousands and thousands of examples. Scientists have collected them from all over the world and laid them out to measure and describe them. Then they go to other museums and compare the tools there. Many scientists from the United States and Europe go to the National Museum in Nairobi to study the tools found by Mary Leakey and her team.

The tools are given a name, according to their style and form. Individual tools have a name such as hand ax, but they are also grouped more broadly into categories denoting the culture or industry.

The early tools that came from Olduvai Gorge are called the Oldowan, which is easy to remember because it sounds like "older one." The Oldowan industry first appeared in levels of the earth that have been dated at 1.8 million years old. The Acheulian style, which is a little younger, is also found at Olduvai. As mentioned before, the Acheulian was named for artifacts from France. The fact that people in Africa and people in Europe had the same style for making their tools is one clue that suggests many of our early ancestors migrated out of Africa to go live in Europe. The more advanced form of the Oldowan tool kit has also been found in Europe, as well as in Java, a large island in Indonesia, and in northern China.

Archaeology was still a young science when Mary Nicol was born in London in 1913. But she would become the most productive archaeologist in Africa, and her skills as an artist would influence her career.

THE INFLUENCE OF AN ARTIST

Mary Nicol began to draw when she was 10 years old. She never attended any art classes, but she must have learned a great deal by watching her father work. Erskine Nicol was a painter, like his father of the same name. Mary's mother, Cecilia Frere, had studied art in Italy. It also happened that Mary's father was interested in archaeology. He took his family to the Dordogne region of France and encouraged Mary to look for flint tools. In Dordogne Mary saw the famous cave paintings drawn by people during the Ice Age, between 35,000 and 10,000 years ago. She also saw the tools and bone artifacts in the museum at Les Eyzies. Her parents became friends with the archaeologist at the museum and visited his excavation site.

Mary and her father searched through the piles of earth after the French archaeologist had already picked out the tools that he wanted. Like Louis Leakey, Mary became a collector of stone tools before she was 13 years old. "For me it was the sheer instinctive joy of collecting, or indeed one could say treasure hunting," she wrote in her autobiography, *Disclosing the Past.* She developed her own way of sorting and classifying her finds. "I remember wondering about the age of the pieces, and the world of their makers."

Mary admired her father a great deal, and they spent afternoons looking for tools or taking long walks. Erskine Nicol was an intelligent man, and Mary was influenced by his many interests, including a love for animals. Mary learned to sit patiently in the woods and watch foxes and wild boars without disturbing them. She probably never dreamed that she might one day live in Africa and be surrounded by wildlife. She was shattered when her father died suddenly in 1926. "I was barely 13, and I had just lost forever the best person in the world."

EDUCATION

Because her father's work had involved so much travel, Mary was rarely in one place long enough to attend a full school term. She did learn to speak French fluently, and a tutor was brought to France to teach Mary history and Latin. After her father's death, she and her mother returned to London, where Mary was enrolled in a convent school. Her impression of school was very similar to Louis Leakey's experience in England. She disliked the formal classroom and found few friends among her classmates, who seemed "utterly juvenile compared to the company I was used to keeping." Mary had been surrounded by sophisticated adults for most of her life.

Mary was so miserable that she pulled several stunts that led to her dismissal from two convent schools. Her formal education ended with a bang, as she describes it. She set off a small explosion during a chemistry lesson. That she had learned enough chemistry to produce a good explosion says that Mary could indeed learn when she was inclined. She learned how to pilot a glider plane, for example, and today she is famous for her discipline and hard, careful work. Once she decided what she wanted to do, she became a very good student.

As a teenager, Mary visited an excavation site in England, where she met an archaeologist named Dorothy Liddell. Mary began to think that archaeology might be the career for her, as Liddell was living proof that the profession was "open" to women. Mary began to attend lectures and courses on archaeology and geology in London, which appealed to her much more than the convent schools. Her mother tried to get her enrolled in Oxford University, but Mary didn't have the formal qualifications, and she was told that she would not be accepted, that it wasn't even worth trying. (In 1981, Mary Leakey was awarded an honorary doctorate from Oxford for her outstanding work in Africa.)

But Mary found out that she could sit in on classes at University College, London, without any qualifications or having to endure the rest of the college requirements. She had been an unhappy teenager, full of rebellion and mischief, but this opportunity marked what she has described as a "return to sanity."

After three years of attending lectures in London museums and the university, Mary began to write letters to various scientists to inquire if she could work at their excavation sites. She received several polite rejections, but in the summer of 1930, she began to work as an assistant to Dorothy Liddell at Hembury in Devon, England. It was a wonderful opportunity for Mary, excavating pottery and other remains from one of the earliest major Neolithic sites of southern England. She also had a good mentor in Dorothy Liddell, whom Mary admired. Their field seasons included the summers of 1930, 1931, 1932 and a spring season in 1934. Between these field seasons, Dr. Liddell asked Mary to draw some of the discoveries, including flint tools. The drawings appeared in Liddell's publications on the Hembury site.

The drawings were so good they attracted the attention of another archaeologist. Dr. Gertrude Caton-Thompson invited Mary to illustrate the stone tools from her excavation work at Fayoum in Egypt. Mary's work appeared in the book *The Desert Fayoum*, and for the first time, she was paid for her drawings. Dr. Caton-Thompson was so pleased with her work that she helped Mary develop her career, inviting her to a lecture at the Royal Anthropological Institute in London. The speaker was a young man from Africa who had made some interesting discoveries at "that remote place Olduvai Gorge" (as Mary described it) and at a site called Kanam, in Kenya.

Louis Leakey sat next to Mary at a dinner following the lecture, and he invited her to illustrate his book *Adam's Ancestors*. Louis was 10 years older than Mary and already an established archaeologist. Yet Mary noticed that he treated her with a great deal of respect, as an equal colleague, which she appreciated. He also loved to travel in wild places, work in the field and observe birds and animals, as did Mary. Louis had already begun to live a life quite separate from his wife, Frida, who did not like his expeditions to Africa. Mary loved nothing better and was already considered a renegade herself.

During the 1934 field season at Hembury, Louis visited Mary at the excavation site, and entertained Dorothy Liddell and her team by making flint tools. Mary later worked at Louis' own site in Swanscombe that summer, which gave her good experience for her own site, called Jaywick. This excavation began in the autumn of 1934, and for the first time, Mary was in charge of her own dig project. She worked alongside geologist Kenneth Oakley, one of the leading scientists of his time. While looking for tools, Mary came across an elephant's tooth that turned out to be the largest ever found in England.* She was slightly embarrassed, however, as she had no idea of what it was when she found it; she had never seen an elephant's tooth before. When Louis arrived on the scene, he was able to identify it immediately. The Jaywick project turned out to be very successful. With Kenneth Oakley, Mary wrote her first scientific paper, which was published in 1937.

A DEBUT

Mary was only 21 years old when she worked at Jaywick, but she had already overcome her lack of formal classroom education, and had worked with some of the leaders in the fields of archaeology and geology. While her report on the Jaywick finds was very good in its own right, it was also a matter of prestige for Mary to publish the paper with Kenneth Oakley, who became associated with the British Museum of Natural History in London. He is perhaps best known for unraveling the famous Piltdown Hoax. While the so-called Piltdown Man is not a legitimate discovery, its influence on the field at the time that Mary was making her debut is important to understand.

---

* An elephant's tooth appears in the photo on page 15.

In 1912, a rather large human-looking skull and a small, apelike jaw were found at a site known as the Piltdown gravels in southern England. Paleontologists put the two pieces together in a reconstruction, even though there were no bones connecting the skull to the jaw. The discovery was proclaimed as a "missing link" and proof that humans had first developed in England. Most scientists were taken in by the discovery, because the large skull suggested that our ancestors were intelligent, even if the rest of the body (in this case, only the apish jaw) was still primitive. So the hoax fit in with the ideas of the day, and for many years the "Piltdown Man" was on display at the British Museum.

As mentioned earlier, in 1925 Raymond Dart reported the Taung child from South Africa. It had just the opposite features of Piltdown—a very small skull and a large, jutting jaw. Dart suggested that the Taung child was one of our ancestors and that humans began in Africa. Many of his ideas turned out to be correct, but they were ignored because most scientists accepted the Piltdown "evidence." Dart's ideas were controversial because of the way he wrote about the find, especially when it came to tool use. Some animal bones had been found in the South African caves, and Dart assumed that these were tools that our ancestors had used in a violent way. Limb bones were seen as clubs, and our ancestors were described as bloodthirsty cannibals. Recent studies in South Africa have shown that most animal bones in caves are actually the leftovers from a leopard's or lion's meal.

Dart eventually did gain credit for his ideas that the Taung child was a hominid, or early human. But this occurred long after the Piltdown Man was exposed as a hoax. In 1950, Kenneth Oakley experimented with fluorine to see if the Piltdown skull was the same age as the other, real fossils found at the site. Fossils absorb fluorine from the earth when they are buried; the amount of fluorine fossils from the same area absorb should be the same—if they are the same age. The Piltdown skull turned out to be a rather modern human skull belonging to a *Homo sapien,* and the jaw was an ape's that had died shortly before it was planted in the gravel at Piltdown. To this day, no one knows who planned the hoax, but it was finally exposed in 1953, 41 years after the bones were found.

In *Adam's Ancestors*, which Louis Leakey published in 1934, he accepted the Piltdown Man as a legitimate discovery, but he felt that his Kanam jaw was a better candidate for a more ancient ancestor. Before Boswell's criticism caused the Kanam jaw to be controversial, Leakey obtained funds for more field research. In October of 1934, Louis Leakey returned to Africa for his fourth East African expedition. He invited geologist Peter Kent to join him, as well as surveyor Sam White, ornithologist Peter Bell, and an archaeologist named Mary Nicol.

MARY GOES TO AFRICA

Mary's mother, Cecilia Nicol, had hoped to persuade her daughter to forget all about Louis Leakey. Mrs. Nicol arranged a tour to South Africa and Zimbabwe (then known as Southern Rhodesia) where they would visit prehistoric sites. They would leave in January, traveling by boat to Cape Town, and return to London in April of 1935—or so Mrs. Nicol had hoped. But Mary had already made plans to join Louis at Olduvai Gorge in April. Her mother inadvertently made the journey easier for Mary, as Zimbabwe is much closer to Tanzania than London.

When they arrived in Cape Town, Mary contacted an archaeologist named John Goodwin, who studied the Stone Age in South Africa. In addition to touring many archaeological sites, Goodwin invited Mary to join him in the field for a few weeks at a place called Oakhurst Shelter, where excavations were continuing inside a cave. Mary was delighted, for she quickly discovered that Goodwin was a very careful and meticulous worker, and she learned much from him. Most of the artifacts belonged to the late Stone Age and came from a level that was dated at under 20,000 years old. The experience proved to be good for Mary's future work in Kenya. In 1939 she found similar tools in a rock shelter near Lake Naivasha.

In Zimbabwe, the Nicols visited the famous rock paintings, which occur in cave shelters and on open cliff faces. The style was quite different from the cave paintings Mary had seen as a girl in France, and they were much more recent. But rather than thinking about the past, Mary was thinking about her future, with Louis.

Seeing the paintings of Zimbabwe made her remember that Louis had promised to show her the rock paintings of the Kondoa area in Tanzania. Her mother's hopes were in vain, and on April 17, she took the train back to Cape Town, alone and in tears. Mary took a plane to Moshi, Tanzania, just south of Mt. Kilimanjaro. Louis arrived the next day to meet her, and together they set out for Olduvai Gorge.

CHAPTER 3 NOTES

p. 25 "a very remote period indeed . . ." John Frere quoted, in Richard Leakey and Roger Lewin, *Origins*, pp. 22–23.

p. 28 "I had been handling stone tools . . ." Mary Leakey, interview with author, 1984.

p. 30 "For me it was the sheer instinctive joy . . ." Mary Leakey, *Disclosing the Past*, p. 25.

p. 31 "I was barely thirteen . . ." Mary Leakey, *Disclosing the Past*, p. 29.

p. 31 "utterly juvenile . . ." Mary Leakey, *Disclosing the Past*, p. 32.

p. 32 "return to sanity . . ." John Reader, *Missing Links*, p. 162.

4

STONES AND BONES

The road to Olduvai is a beautiful introduction to East Africa. It travels south from Arusha to cross the Great Rift Valley near Lake Manyara, then up and around the spectacular rim of Ngorongoro Crater. When Louis and Mary made their journey in 1935, they saw more of this steep section of the road than they had planned, for they had to push their car up the volcanic slope.

Today the Ngorongoro Conservation Area is one of the most popular safari destinations in Africa. Two million years ago, the cone collapsed in the center, creating a caldera. Most calderas around the world fill with rainwater, creating vast lakes, but this one is full of animals, discouraged from migrating because of the steep slopes that stretch for 2,000 feet from the inner floor to the rim. The floor, about 12 miles in diameter, contains rare black rhinos, hippos and many lions. On their return trip, Louis would compare the hippos in the caldera to the ancient hippos he and Mary found at Olduvai.

But on their inward journey, Ngorongoro became an obstacle, with heavy rains turning the dirt road into thick, slippery mud. It took them two and a half days to make 16 miles up "that impossible slope," as Mary described it, and they must have been miserable, because they ran out of food and dry clothes.

It was a classic case of push and struggle, with beauty just on the other side of the hill, which also describes their research at Olduvai. They wouldn't see the "top of the hill" in research at Olduvai until 1959, when Mary made a great discovery that changed their lives and put them into the pages of *National Geographic* magazine. Now, of course, they were changing their lives too. They would get married in 1936, and this expedition to Olduvai was the beginning of a special partnership. It would take 25 years for the rest of the world to notice the results of their efforts, but scientific achievements often occur at the end of a long, meandering path of tedious, even boring, work. In Africa, the tedium is often relieved by pure adventure.

Once they got to the rim of Ngorongoro, the road swept them down toward the Serengeti plains, with a spectacular view of volcanoes in the distance. It was a road lined with wildflowers, and they saw herds of giraffe that fed upon acacia trees. Maasai in traditional robes the color of red ocher herded their cattle in the distance. That's the pretty part.

WORKING AT OLDUVAI

Olduvai Gorge itself is a hot, arid, desert landscape. Louis described his first expedition there as unpleasant; it was difficult to find water, which had to be rationed. There was also a terrific wind that wouldn't stop blowing.

> *it carried with it vast quantities of fine black dust that filled every corner of the tents and could not even be kept out of the mess tent. If you spread some semi-liquid sun-melted butter on a piece of bread it would be covered with a fine black dust before you could get it into your mouth. If you poured out a cup of tea or coffee in a few minutes it had a fine black scum of dust on its surface. You breathed dust-laden air, your nostrils were filled with dust, you ate dust, drank dust. . . . The heat of the sun was terrific. . . . and the dust mingled with the sweat to make your body filthy. And yet water was so scarce.*

This first visit was full of adventures for Mary; she met a rhino when walking up a narrow path and almost stepped on a lioness that slept in the tall grass. But with only three months allotted for their work, the team got down to business fast. As they began to find fossils, Louis named the various sites after people; one site was named MNK, for Mary Nicol Korongo—*Korongo* means "gully" in Swahili. There, toward the end of the expedition, Mary found two fragments of a hominid skull, which were classified as *Homo erectus*, the large-brained toolmaker. *Homo erectus* lived at

Olduvai as early as 1.6 million years ago. They also found hand axes lying nearby as well as the remains of antelope and pigs.

The great promise of Olduvai was the fact that there fossils of early humans were often found in the same bed, or layer, as their tools. Naturally, it made one think that these were the early humans who used these tools. Perhaps the antelope and pig bones were the remains of one of our ancestors' campsites. All of the evidence would have to be studied carefully and then compared with evidence from other sites. At the end of the expedition, Louis and Mary packed all of the stones and bones into wooden boxes, padding their finds with dry grass, for their return journey to England.

Olduvai was a dream come rue for anyone interested in the search for human origins, because the levels of the area, or geological beds, revealed several different eras of human evolution. Mary Leakey and geologist Dick Hay divided up these levels into five beds, for reference. Bed I, for example, was the oldest, while Bed V was the most recent. They would eventually uncover evidence of a toolmaker younger than *Homo erectus*. In fact, it would be named *Homo habilis*, which means "Handy Man." It would come from Bed I and be dated at 2 million years old.

From the beginning, when Professor Kattwinkel "fell" into his discovery, it was clear that Olduvai held great potential. Now the Leakey team began to expand that potential even further, exploring a new site nearby. Their tip came from a keen-eyed African who lived in the area.

At every field camp, Louis Leakey set up a small clinic for local Africans, who had a lot of trouble with malaria, spear wounds and flies infecting their eyes. Many Africans would go to the clinics in hopes of help, because there were no hospitals for hundreds of miles; others would visit just out of curiosity. One visitor was a man named Sanimu, who looked at their collection of fossils and promptly said the he too had noticed "bones like stone" in an area to the south, toward Lake Eyasi.

Sanimu set out on foot to gather a few samples. When he returned, Louis and Mary were delighted to see that what he held in his hands were actually real fossils and a few fossilized teeth. They immediately made plans to visit the area, called Laetoli.

Years later the site would produce one of Mary's greatest discoveries, the famous footprints (see p. 3 and, pp. 98–99) which proved to be the earliest evidence of upright walking by humans, as well as some of the oldest hominid fossils ever found in Africa.

THE LONG ROAD TO THE TOP OF THE HILL

If fossil finding appears to be so easy, why did it take the Leakeys so many years to make great discoveries? For one thing, they spent their early years locating many sites and making preliminary, surface digs. Most of these sites were so full of stones and bones that they are still being studied today. And the Leakeys had only a small grant for limited research. Like many paleoanthropologists, they had to work at other jobs to make a living, and do their research part time.

Progress was painstakingly slow. Louis and Mary first went to Laetoli in 1935, but Mary would not return there again until 24 years later, and she would not begin a major excavation there until 40 years later! These delays show nothing if not patience. There was so much to be found, so many promising sites in Africa, and there were very few people who could do the work.

The Leakeys employed other scientists and Africans to help them search for fossils, but once fossils were found, they would have to supervise a tedious and careful excavation not just of a particular fossil but of the entire surrounding area, often digging a dozen yards square.

To learn anything from a hominid bone, its context—where it was found in relation to other bones and artifacts, as well as its position in the earth—must be studied. This meant finding examples of the other animals that lived at the same time, and carefully counting and measuring bones and stones that were found in the same area. The position of every fossil or tool that was lifted from the ground had to be recorded and drawn on an accurate chart. A grid system had to be developed for every site, measured off in square meters (each one slightly larger than a square yard). The geological level of the discovery had to be determined; this meant digging trenches into the earth. Trenches also can tell you whether stones and bones have been washed there by a flood, or if they are truly part of the natural deposition of the landscape. Paleontologists often refer to the earth as "deposits," because the upper layers have been deposited by the winds and the rains. A level of a few feet may take several thousand years to deposit. The process of deposition is not something that we see happening within our lifetimes, except in the desert, where sand dunes visibly move from year to year. If an overflowing river cuts through the earth, layers of soil may be eroded and the stones and bones may shift. Similar movement happens with the rise and fall of a lake. Both of these movements of water had occurred in the history of the Olduvai area, and the Leakeys had to be careful in making assumptions about what they found.

By the end of their 1935 survey, they had found 30 sites at Olduvai Gorge. Eventually there would be 127 sites.

Geologists were always brought in to date the sites, and their work also relies on a good fossil collection. Fossils as old as those at Olduvai are impossible to date directly, so geologists estimate an age based on their position between older and younger levels of earth, which contain volcanic rocks and ash that can be dated directly.

The ages of fossils of fauna in the area are then compared with these estimates. Faunal fossils are much more abundant and thus provide more information than hominid fossils. A certain species of pig that evolved very rapidly has provided a great deal of evidence on a site's age. The stages of the pig's evolution can be determined by the shape of its teeth, and pigs with a certain shape tooth have been dated at certain ages.

Olduvai would eventually become one of the best-dated sites in Africa. When on archaeological digs, *everything* can be important. To make certain that nothing is missed, every bit of soil is carefully sieved through a screen. The sieve actually looks like a window screen, with a thick wooden frame on the sides and handles at each end. Two members of the team shake the dirt back and forth through mesh fine enough to stop a single mosquito. Then they look through the remains, recovering small bits of bones, even rodent teeth. For example, from one site at Olduvai, Mary Leakey and her team excavated 55,000 square feet. A total of 37,127 artifacts and 32,378 fossils were recorded—that second figure does not include over 14,000 rodent fossils, plus fragmentary finds of birds and frogs!

Even though the Leakeys had plenty of work to do at Olduvai, they were intrigued by the fossil that Sanimu had brought to them and could not resist a closer look at this new site.

EXPLORING LAETOLI AND KONDOA

With Sanimu as their guide, Louis and Mary made the journey to Laetoli. The Laetoli area was thick in volcanic ash, and they found many fossils that were well preserved. From the very beginning they were intrigued by the Laetoli site, but they only had three months to cover a lot of ground in Tanzania.

By the time they returned to Olduvai, the water and food supplies were low, so they sent a truck to Nairobi for supplies. For two weeks they continued to work, surviving only on rice, sardines and apricot jam. Finally they set out to find out what had happened to the supply truck. Then they had an accident.

Near the Kenya border, Louis drove off the road into a deep gully. Neither he nor Mary was hurt, and the vehicle was not damaged, but they were forced to spend the night there. The next day, they had to figure out how to get the car back up on the road. They decided to dig a trench, but they had nothing to dig with but plates and table knives. From dawn until dusk they dug, finally completing a good trench, only to look up and see the truck from Nairobi drive up, with all the supplies, including shovels.

After a brief return to Olduvai, they went to see the rock paintings of the Kondoa district, as Louis had promised. Sanimu was invited to go along as their guide, which was typical of the way the Leakeys worked. Although untrained in the world of science, Sanimu had proved his skills of observation, noticing things on the landscape. Over the years the Leakeys would employ a team of talented Africans, and several of them would make great fossil discoveries.

On their journey to Kondoa, Sanimu's ability to speak the local Maasai language was very useful in locating remote and unexplored rock paintings. Mary was entranced by the drawings: simple and beautiful paintings of human stick figures—one man plays a flute, with music "dripping" from the tip of his instrument. There were white giraffes and ostriches. Many of the paintings were on top of others, suggesting that the "rock canvas" had been used for many years, with people painting one figure over another. Mary vowed that she would return some day and spend more time excavating around the sites, to see if she could find any traces of the people or their artifacts. Here, too, is another example of how long it takes for this sort of research to come to much. Mary would return to Kondoa 16 years after her first visit. In a way, it must have been painful to see all these things and imagine what secrets they held. On the other hand, it was wonderful to know that Africa held such promise. After all, this was Louis' fourth official expedition to Africa, surely there would be a fifth. As it happened, the next official archaeological expedition would be Mary's to lead.

RETURN TO ENGLAND

In the summer of 1935, Louis and Mary returned to England, to live in a country village at a place called Steen Cottage. It was there that Louis wrote his first autobiography, *White African.* A lecture series of his was published in a second book, *Stone Age Africa,* which Mary illustrated. These provided their only sources of income, and they lived simply but happily—for the most part. Mary's mother still disapproved of her relationship with Louis, and his divorce from his first wife would not be completed until 1936. But at the end of 1936, things began to look better.

Louis received a grant to prepare a study of the Kikuyu. It would complicate life, for they would have to return to Kenya and give up their cottage and the dogs that they had collected as pets. But as Mary notes, "our philosophy was not to fret over difficulties but to accept opportunities and overcome obstacles as they presented themselves." On Christmas Eve of 1936, Louis and Mary were married, and in January 1937 they set out for Kenya.

While Louis worked on his Kikuyu studies, Mary began an excavation two miles south of the town of Nakuru. The place is called Hyrax Hill, after the furry and very vocal mammals, about the size of rabbits, that live there. In July of 1937, the Leakeys set up a tented camp and a work hut. Louis brought along two elders from the Kikuyu tribe, whom he planned to interview at length, and Mary began to work on this burial site, which had been a Late Stone Age settlement. She unearthed many stone tools made of obsidian as well as pottery. On a higher level of beds, she also found a more recent Iron Age settlement, which included a few iron objects, glass beads and bones of domestic cattle. Local citizens became interested in the site and set up a fund to help finance Mary's work. One neighbor was so impressed that she bought the land and gave it to the government, to create a small center for visitors and protect the area. Today Hyrax Hill is one of the many archaeological sites that people can visit in Kenya.

It was at Hyrax Hill that Mary got her first Dalmatian in Africa; these sleek black-and-white dogs became a part of her life from then on and accompanied her during all her years of research at Olduvai, becoming something of a trademark.

In 1937 and 1938, the Leakeys worked at a Late Stone Age site known as Njoro River Cave, which was located on the farm of Nellie Grant. Mrs. Grant was the mother of author Elspeth Huxley, who wrote *The Flame Trees of Thika*. Her mother appeared in that story as Tilly. Mrs. Grant had visited the Leakeys at the Hyrax Hill site and said she thought she had a similar place on her own farm. The site turned out to be important, with unique evidence of cremation. The Leakeys unearthed 80 individuals and were able to tell that the people had been bound tightly and wrapped in skins before they were burned. Carved wooden vessels were among the artifacts, as were the remains of gourds and pottery. The finds were dated at about 900 B.C.

When Louis finally finished his report on the Kikuyu, it was 700,000 words long. No one wanted to publish such a large volume of work, and Louis refused to edit it down. (It was finally edited and published in 1977 by Academic Press, with funding from the L. S. B. Leakey Foundation, now based in Oakland, California, several years after Louis Leakey's death.)

When World War II began, the British government hired Louis as an intelligence officer. His travels around Kenya and Ethiopia in that capacity gave him an opportunity to look for new fossil sites, and he located a rich area just north of the Kenya/Ethiopia border, in the Omo River Valley. The Omo River flows into Kenya's Lake Turkana.

In 1940, Louis became the curator of the colonial museum in Nairobi, which was then known as the Coryndon Museum. The Leakeys moved into a small house on the museum complex, a bungalow with a tin roof that today serves as the headquarters for the Wildlife Clubs of Kenya, a model organization that offers students an opportunity to learn about conservation and natural history. Louis took on the job at the museum in addition to his wartime duties and incessant hunts for fossil sites. Meanwhile Mary began to work at yet another excavation site near Lake Naivasha, unearthing an enormous number of stone tools and flakes. From one trench alone, 75,000 artifacts were recovered. The material was stored in cardboard boxes, and before Mary and her team had a chance to sort it all out, termites ate the boxes and the whole collection fell into a jumble. It was such a mess that Mary didn't have the time to try to sort it out. Mary was also expecting their first child; Jonathan Leakey was born in November of 1940. Today Jonathan is best known for his collections of venomous snakes, which he milks for antivenom, but as a boy he contributed to one of the Leakeys' most important discoveries at Olduvai.

Mary did not let parenthood stop her work, and four months after Jonathan's birth, she left him in the care of Louis and a nursemaid and set out for an expedition that would take her back to Tanzania and the Ngorongoro Crater. Her traveling companions were Joy and Peter Bally; Joy is remembered today for her work as a conservationist with her second husband, George Adamson, and her book on their life with lions, *Born Free*. Mary did a small excavation on the floor of the crater, with a grant from Cambridge University of less than $100. She unearthed burial mounds similar to those at Kenya's Njoro River cave.

It was a busy time for the Leakeys, who were virtually surrounded by new sites to explore. One of the most interesting sites is only an hour's drive south of Nairobi. Today it is remarkable to go to Olorgesailie (Ol-or-ga-′sa-lee) and, within such a brief time, move from modern hotels and city traffic to a landscape full of hand axes made by our ancestors. Nairobi is the capital of the fastest-growing nation in the world, and it is fascinating to think that within these two settings, you can see our past and our future.

OLORGESAILIE

Olorgesailie is located in the eastern branch of the Great Rift Valley. It is only a few miles south of the Ngong Hills, which were made famous in Isak Dinesen's book *Out of Africa*. (Dinesen is the pseudonym of Karen Blixen.) On the southern side of these hills the earth drops down dramatically to the valley floor. Several volcanic cones rise up in the distance, and by studying the landscape closely, you can trace several red fault lines that mark the areas where the earth has shifted. The site was first noticed by geologist J. W. Gregory, who figured out how the Rift Valley was formed in the 1890s. While studying geology, he noticed stone tools, which he wrote about in his report on the area. On Easter

Many hand axes lay where they were found at Olorgesailie, just an hour's drive south of Nairobi. Louis and Mary Leakey found hundreds of these tools there in 1942. Today the site is open to the public, with a small museum and camping facilities. (© Delta Willis)

weekend in 1942, the Leakeys and a couple of friends drove down to see if they could find any of these tools.

The basin that they explored was once part of a lake, a mile in diameter. This basin is covered in diatomites, the tiny white silica shells of algae that flamingos feed on. The algae that the flamingos did not eat died naturally and left only their shells, or skeletons, behind. So when Lake Olorgesailie dried up half a million years ago, it left an area covered in diatomites.

The Leakeys and their friends spread out, each of them moving in a different direction. Louis and Mary came upon stone tools at the same moment, and shouted to each other. They both saw hundreds of tools, just laying there on the surface. Over the next 10 years, more than a ton of stone tools was found.

The tools found at Olorgesailie are Acheulian and include many hand axes and cleavers. Their age is between 600,000 and 800,000 years old. For several months in 1943, Mary carefully recorded the many discoveries. She camped out near the site with young Jonathan, while Louis continued to work in Nairobi at the museum. At night, in her tent, Mary heard lions roar nearby.

Today there is a small museum at the site, with an elevated boardwalk built over the excavation areas, where you can see the tools and some of the trenches that Mary made to study the different levels. When asked what it looked like when they were first discovered, Mary replied, "Just as you see it today. All of those tools were there together, just hundreds of them."

Several archaeologists came to study these rich assemblages of tools, including the late Glynn Isaac, who wrote about the sites. At first it was thought that these tools marked campsites. Today new research suggests that the tools might have been washed up there in a flood. That doesn't mean that early humans didn't make them; it simply means that humans may have brought them there and left them, or the tools may have washed up there from a place where hominids stored them. So far, no hominid bones have been found in the Olorgesailie basin itself. But in 1986, the bones of an elephant were discovered by Rick Potts, who is now following up on the research begun by the Leakeys. Potts and his team made casts of the elephant bones, which are featured in an exhibit on Olorgesailie at the Smithsonian Institution in Washington, D.C.

From this relatively young site, the Leakeys would begin to investigate an area rich in Miocene-era fossils, 16 million years old. The place is called Rusinga Island, which is located near the eastern shores of Lake Victoria. Rusinga Island is so rich in important fossils that many are still being found there today. At Rusinga, Mary would find a skull that belonged to our oldest known ancestor.

CHAPTER 4 NOTES

p. 37 "that impossible slope . . ." Mary Leakey, *Disclosing the Past*, p. 54.

p. 38 "it carried with it vast quantities . . ." Louis Leakey, *White African*, p. 296.

p. 43 "our philosophy was not to fret . . ." Mary Leakey, *Disclosing the Past*, p. 67.

p. 47 "Just as you see it today . . ." Mary Leakey, interview with the author, 1984.

5

PROCONSUL

The Leakeys made many journeys to Rusinga Island between 1942 and the 1950s. Richard, who was born in 1944, remembers the family's month-long stays as exciting times, anticipated by careful planning, going over the list of supplies they would need. "For several weeks before setting off," he recalled, "we all felt we were preparing for an adventure. . . . Plans were made, and on the day of departure we would be awake long before dawn."

Their supplies already had been packed into wooden boxes and loaded into the car the day before. Blankets and bedding were thrown over the top of the luggage, and the Leakey boys rode on top of this. By sunrise they were on their way to Naivasha and then Nakuru, both Rift Valley towns to the west of Nairobi where they made stops for fuel. After a picnic lunch near Nakuru, they began the long, bumpy journey to Kisumu, on the northeastern shores of Lake Victoria.

There they boarded a boat for Rusinga. At first they used an Arab dhow, a type of sailboat. But eventually, with new funds for research, Louis Leakey bought a 44-foot cabin cruiser, which they named the *Miocene Lady* in honor of the time frame, or geological era, of the fossils they found. Unlike Olduvai Gorge, where the erosion features different exposures between 1 and 2 million years old, most of the exposures on Rusinga were from 14 to 20 million years old.*

It was a long day's drive, and by the time they arrived at Kisumu port to board the boat for Rusinga, it was dark. All of the supplies were loaded onboard.

---

* See the geological time scale on page 8.

Louis liked to complete the seven-hour cruise before midmorning the next day, to avoid the rough waters on Lake Victoria that began as the sun heated up the wind currents. So they traveled over dark waters, and Richard remembers watching shooting stars before dawn. As they approached Rusinga Island at sunrise, they encountered many local Luo fishermen, paddling long wooden canoes. They bought fresh fish for breakfast and then moored the boat in a bay near the fossil site.

The Leakeys slept onboard the *Miocene Lady* during several of the expeditions; tented camps were set up on land for other visiting scientists and staff, as was a work tent. The work tent had tables for studying fossils, storage boxes, and the equipment needed in the field—dental picks, brushes, glue and toilet tissue for wrapping and protecting fossils when they had to be carried.

Before everyone's daily bath in the lake, Louis would fire a couple of blasts from his shotgun, to frighten away the crocodiles. Actually the many fish in the lake provided abundant food for the crocodiles, which rarely bothered people.

While mornings were spent excavating known sites, during the cooler, late afternoons, Louis walked around the island looking for new fossil sites. Jonathan and Richard went along on these walks. Besides looking for fossils, the boys learned a great deal about the island's birds and natural history. The boys developed an early interest in butterflies and beetles, and began to collect them.

DISCOVERING MIOCENE BONES

The Leakeys were not the first people to search for fossils on Rusinga. In 1927, H. L. Gordon discovered some apish-looking bones at a site called Koru. The finds, which included a few jaws, teeth, and limb bones, were studied by British paleontologist A. T. Hopwood, who believed they belonged to an ancestor of the chimpanzee. So Hopwood named the new genus *Proconsul*, meaning "before Consul." ("Consul" was a popular trained chimp that performed in London vaudeville shows at the time; he wore a cap and a jacket, rode a bicycle and smoked a pipe.)

Proconsul is classified into a group of apes known as the dryopithecines, or "woodland" apes. Their wrist bones are like a monkey's, which suggests that *Proconsul* was a tree dweller. (This makes sense, for when *Proconsul* lived, unlike now, there were forests on Rusinga Island.) Their chest was broader than a monkey's, and they had no tail, but a vestigial tailbone (an evolutionary remnant) like our own. The dryopithecines disappeared about 9 million years ago. The first *Proconsul* bones that were discovered are about 18 million years old. Based on the bones of this animal, it is thought that some of their population evolved into other species at 9 million years ago. Hopwood believed that this was the ancestor of modern-day chimpanzees.

At the time, it was thought that the human ancestral line went back into the early Miocene. If so, Rusinga was promising ground for finding an in-between form of humans and apes. Louis Leakey, keen to discover such a missing link, became especially interested in the area when he found some apish-looking fossils himself during his second expedition to Rusinga in 1932. He also found a way to create a great deal of interest in the potential of the area.

THE FIRST CONFERENCE

In 1946, the Leakeys began to organize the First Pan-African Congress of Prehistory and Paleontology. It would be held in Nairobi in January of 1947. The idea for the conference, as Mary recalls, "was very much Louis' brainchild."

He wrote letters to leading scientists from all over the world and even visited some in England and France, convincing them to travel to Nairobi, which was an expensive trip. Organizing the conference required a tremendous amount of effort, but bit by bit, the Leakeys pulled it together. Raymond Dart, the discoverer of the Taung child in South Africa, agreed to attend, as well as his colleague Robert Broom, who did much to prove Dart right by finding further evidence of australopithecines in South Africa. The great French expert on prehistory, Abbé Henri Breuil, was named the president of the conference, adding to the prestige of the event.

British geologist Kenneth Oakley, with whom Mary had published her first scientific paper, would be there. England's leading expert on primatology, Wilfred Le Gros Clark, agreed to go, as well as paleontologist Desmond Clark, who had discovered several fossil skulls in Zimbabwe. It was a monumental gathering, a historical meeting of the minds.

The conference itself was held in the Nairobi Town Hall. The participants drove down to Olorgesailie, where a special ceremony marked the official opening of the outdoor museum near the stone tools discovered in 1942. Mary Leakey led the visiting scientists on a tour of Hyrax Hill, and several other sites in Kenya were visited. Then the group of 50 people set out on an extensive safari to Tanzania, exploring Olduvai Gorge, Ngorongoro and the rock paintings near Kondoa. Even today, organizing a safari for 50 people to such remote areas is a complicated task. But most of these scientists were used to working in the field, and even though travel in remote parts of Africa presented special problems, it proved to be a great adventure.

Afterward, a small group of scientists stayed on to visit the Rusinga Island sites. Just before they reached the island, their boat ran onto a submerged rock. The Luo men of the island promptly waded out to help carry the passengers ashore; Mary Leakey remembers that one elderly lady was so shocked that the Luo were naked, she covered her eyes. But rather than wade with the crocodiles, the lady accepted a ride on the shoulders of one man.

While only a few scientists traveled to the Rusinga sites, Louis had carefully arranged an exhibit of the fossil discoveries from the area at the museum, which everyone saw during the conference. Several scientists also gave reports on the importance of the finds from Rusinga. By this time, more examples of *Proconsul* had been found, as well as fossil mammals, reptiles, and fish. This impressed everyone, including Le Gros Clark, who was inspired to join the Leakeys in obtaining support from the Royal Society for more research at Rusinga. The project was called the British-Kenya Miocene Expedition, and research was planned for 1947 and 1948.

The idea for the conference, and the careful promotion given the Rusinga site, is a good example of how Louis Leakey succeeded in inspiring people and obtaining funding. Mary also arranged several dinner parties and spoke before the group about her finds at Hyrax Hill. Mary, who was not as accustomed to the spotlight as Louis, took up public lecture tours only a few years ago. She was never interested in promoting herself the way Louis was. However, as long as she could speak about the discoveries, this helped her to focus her insight and enthusiasm for the findings.

The meeting itself offered a place to exchange ideas, but perhaps the greatest benefit was future research at Rusinga. Leakey's plan of interesting a wider group of scientists proved to be a profound success.

The 1947 field expedition produced more fossil apes as well as a skeleton of a previously unknown species of rhino. Louis wrote a letter to *The Times* of London about these discoveries and how urgently funds were needed to continue this kind of research. An American businessman named Charles Boise read Leakey's letter and sent a donation of £1,000, which was a lot of money in those days. The donation made a huge difference in the Leakeys' overland journeys; they brought a big truck, which they converted into a mobile home, with bunk beds and cupboards. Boise continued to be their supporter for several years and eventually visited the Leakeys in Africa.

FINDING THE FIRST "MISSING LINK"

In September of 1948, the Leakeys returned to Rusinga Island, camping on the west side near a site called Kathwanga. Within a few days Louis found the bones of an extinct species of crocodile and began to excavate it. Mary didn't care for crocodiles (either "living or fossil," she notes), and she began to wander around in the area, which was eroded and might reveal fossils that did interest her. Of course new species of crocs were important to the general fossil record, but to Mary, the "real prizes among all these Miocene creatures, so far as I could see, were the fossil apes, among which we might hope to find some evidence of man's own line of evolution."

Mary didn't wander around for very long before she noticed several bone fragments from a skull and then a tooth. She thought that it might be exactly what she had been looking for. Within a few minutes of finding more pieces, she shouted for Louis, who came running.

Together they carefully brushed the sediments away from the single tooth, which they recognized as belong to *Proconsul.* As they continued to uncover the find, they found that the tooth was still in place in the jaw. This was exciting. Teeth often pop up by themselves, without the jaw. Knowing how the teeth are arranged in the jaw is important, because as humans evolved, the jaw shape changed from a V shape to a U shape. In the end, Louis and Mary found all the teeth as well as both the upper and lower jaws.

The excavation of the *Proconsul* skull went on for several days, with careful sieving of the surrounding earth to recover even small bits of bone. The Leakeys' top assistant, Heselon Mukiri, was in charge of the sieving operation, sifting dirt through a mesh screen and examining every small bit to see if it might be a bone. Mary spent long hours trying to piece together the 30 or more separate fragments of the skull. Once she dropped a small piece on the dirt floor of the tent and spent a lot of time looking for it on her hands and knees. Even though it was a small bit, it was a key link that joined two larger pieces, and she would not give up until she found it.

As they began to study the many fragments, the Leakeys saw that they had parts of the face and more than half of the skull. A good skull of *Proconsul* had never been found before, and Mary Leakey, who is a very restrained and modest scientist, described the discovery as "wildly exciting."

Having enough of the skull to estimate the brain size of *Proconsul* was essential to the study of human evolution. One of the unique aspects of human evolution, beyond our upright stance, is the growth and evolution of the human brain through time. For many years, the volume of the actual brain size was measured as a way of showing increased intelligence. For example, a skull was filled with water or seeds, and that volume was compared to that of skulls of other species. More recently scientists have focused on the organizational changes within the brain, such as the capacity for speech. Evidence for this can be discovered in later finds by a hollow on the inside of the skull. They also compare brain size to body size, because as primates go, humans have large brains compared to the rest of their bodies.

Mary Leakey's 1948 discovery would create a great deal of interest among colleagues, and its importance is still talked about today. *Proconsul* is no longer thought of as an ancestor to the chimp but rather as one of our shared ancestors.

THE LONG ROAD TO DISCOVERY

Back in Nairobi, Louis decided that Mary should be the one to take the discovery to England for study. To cut her travel costs, he arranged with British Airways (then BOAC) to give her a seat in exchange for publicity.

With the skull in a box on her knees, Mary Leakey set out for London in a converted bomber plane. When she disembarked for a layover in Cairo, the captain locked *Proconsul* in the cockpit. After a press conference at the London airport, she carried the box to Paddington Station, where she took the train to Oxford. There Mary handed over her important carry-on luggage to Le Gros Clark, who studied the skull and drew a graphic reconstruction.

The skull had been crushed and was lopsided. This often happens, because the pressure from the sediments or ash in which skulls are buried causes them to collapse. The brain rarely fossilizes, so the skull is an empty "ball" of nonflexible material that moves and cracks under pressure.

To draw a graphic picture of the find, Le Gros Clark, considered the expert in reconstruction, had to imagine how the skull looked when the creature was alive. This is not totally farfetched, because he was an expert on primates, and the bones of the skull tend to fit together in a pattern. There are grooves that help identify many pieces, and some parts of the skull are thicker than others. The brow ridge above the eye, or orbit, for example, is distinct from other bones. Reassembling a skull is like piecing together a broken coffee mug—but with a coffee mug, you have all the pieces.

However, because all the bits to the skull, including several key back pieces, were not found, Le Gros Clark had to guess at some of the positions and curves. The discovery of a new bone often changes the shape of the "guesstimated" skull. This is exactly what happened with *Proconsul*, but it would take several decades to figure this out.

TEAMWORK

While this book is a story about the Leakey family, it is important to remember that science is a team effort. Even though the Leakeys were leaders and had innovative ideas, part of their success depended on the work of many individuals. The African named Sanimu, for example, led them to Laetoli and helped their research at the Kondoa rock paintings. His compatriot Heselon Mukiri not only did the sieving for *Proconsul* bones but helped Mary Leakey excavate at Olorgesailie; he also found fossils for Louis Leakey in the Omo River Valley. Eventually these African fossil finders grew into a team known as the Hominid Gang, because they were so successful at finding what the Leakeys were looking for.

In addition to this team, graduate students and scientists often went to Africa and worked alongside the Leakeys. Other scientists such as Le Gros Clark, formed vital parts of the team but contributed from a laboratory on another continent. Over the years, the team grew into a great international effort, largely based on respect and personal friendship. Most of today's leaders in the field were at one time or another part of the Leakey team, or befriended by them.

Martin Pickford, one of Richard Leakey's school friends from Kenya, found missing pieces to the *Proconsul* skull—33 years after Mary found the original jaws and face. The discovery is a long and involved story that shows how complicated this kind of work is.

PIECING TOGETHER PROCONSUL

Martin Pickford has found fossils in Pakistan, Namibia and the Samburu Hills of Kenya. But he made one of his most important discoveries in the comfort of the archives of the Nairobi Museum.

In 1981 Pickford was reading through Louis Leakey's notebook on the 1947 expedition to Rusinga. Field notebooks are like detailed diaries. Pickford noticed an entry for "possible primate" bones found at the Kathwanga site. He knew that any primate bones from that area might be important, because Mary had found the *Proconsul* skull there. So he began to dig around in the drawers of fossils from that site. He became intrigued by a couple of bones that were labeled "turtles." He picked them up and turned them around slowly in the light. He recognized them as parts of a primate skull. They had simply been mislabeled.

Pickford took the two pieces of bone to Alan Walker, a British anatomist who also worked at the museum. Walker is an expert at reconstructing skulls, and at the time he was working on making an accurate model of the *Proconsul* skull, which had a gap at the very back. Amazingly, the two bits of "turtle" fit perfectly into the back of the skull.

First of all, how could this have happened? The two men looked at photographs taken during the 1948 expedition. They saw that the skull was lying face down. The two bits from the back had been compressed by burial and had been washed or eroded to the surface in 1947. While Louis Leakey had considered them "possible" primate bones, for some reason they had been mislabeled as "turtles."

Second, what does it all mean? Alan Walker and his team wanted to minimize all distortion and construct an accurate model of the way *Proconsul*'s skull was shaped when it was alive. To do this, they had to make a cast. Then they scored and broke the cast, and reassembled it. They put a mirror to the accurate, full side of the skull and, by using latex to fill in the other side, made a symmetrical match. Then they inserted the two "turtle" bits into the back, and the whole shape of the skull had changed; it was a different-looking animal from the one Le Gros Clark had reconstructed. They discovered that *Proconsul* had a much larger brain that anyone had thought.

To prove that this was significant, they needed a whole skeleton, to compare the brain size to the body size. They decided to go back to Rusinga and search for more bones. No one had been to the site in 30 years, but they found the geologist's notes. Once at Rusinga, Walker stood on a hill reading from the notes while Pickford paced off his directions. Finally Pickford shouted, "I'm here, but I don't see anything." From where Walker was standing, he could see plenty. Pickford was standing in a patch of green grass, in the middle of brown grass. The excavation site had become a little hollow that held rainwater, creating green grass. The two men found so many fossils on the first day that they couldn't carry them.

Walker returned with a team and equipment the next year, and recovered 10 tons of rock, rich in fossils. From this they were able to put together not only a complete skeleton but they found nine other *Proconsul* skeletons. It is one of the richest collections of one species ever made in the search for human origins.

Now the evidence suggests that *Proconsul* was not an ancestor to the chimpanzee alone but a common ancestor that led to all the great apes and humans. If you imagine the branches of a tree, *Proconsul* is on the older main stem, and two forks from that branched into the great apes on one side and the hominids on the other. Mary Leakey found evidence of our oldest known ancestor in 1948, but it took over three decades to figure this out.

With Mary Leakey's help, Walker is still uncovering and assembling the finds from Rusinga today. His team has uncovered bones of infant and juvenile *Proconsuls* as well as a rare skeleton of the antelope known as the water chevrotain. The volcanic ash and mud was so rich on Rusinga that it preserved extraordinary details—such as fingertips and even paths where nerves had gone into the earbone. Among Louis and Mary's 1947/48 finds, there is a bird with distinct impressions of little feathers on its breast and fossil grasshoppers with legs and tiny antennae. They even found a lizard with its tongue hanging out—the flesh of the tongue was preserved. "One day," Walker says, "we're going to turn over a rock and find a *Proconsul* face."

However, the face that the Leakeys found after Rusinga was a giant compared to *Proconsul*. It looked like Darth Vader on a bad day. It had huge jaws, bold jutting ridges above its eyes (or orbits), a crest on top of its head and teeth so big that the press called it "The Nutcracker Man."

CHAPTER 5 NOTES

p. 49 "For several weeks before setting off . . ." Richard Leakey, *One Life*, p. 21.

p. 51 "was very much Louis' brainchild . . ." Mary Leakey, *Disclosing the Past*, p. 91.

p. 53 "living or fossil" crocodiles, Mary Leakey, *Disclosing the Past*, p. 98.

p. 54 "wildly exciting . . ." M. Leakey, *Disclosing the Past*, p. 98.

p. 57 "I'm here, but . . ." Martin Pickford, as quoted in Delta Willis, *The Hominid Gang*, p. 123.

p. 58 "One day . . ." Alan Walker, as quoted in Delta Willis, *The Hominid Gang*, p. 125.

6
"ZINJ"

The 1950s were a decade of change for the Leakeys. Until then their research depended on small grants, plus the fees and royalties Louis earned writing books. The continuing support of American businessman Charles Boise and the Wenner-Gren Foundation in New York allowed them to return to sites in Tanzania. Now they had the money to excavate, yet Louis had very little time to devote to field research in Tanzania. Working at the Nairobi Museum, he was an employee of the government of Kenya, which at the time was officially still a British colony named British East Africa. The 1950s brought powerful conflicts in Leakey's stance as a "White African," because the "whites" and the "Africans" came into direct conflict.

In the same decade that the U.S. Supreme Court ruled that separate was not equal and racial desegregation of public schools began, the so-called Mau-Mau* rebellion of 1952 to 1956 began in Kenya, led by a Kikuyu named Jomo Kenyatta. Kenyatta not only spoke the same languages as Leakey, but he had attended the London School of Economics, and applied what he learned to real estate. The native Kenyans' struggle to retain rights to their land found Louis Leakey, a member of the Kikuyu tribe, broadcasting propaganda for the British on the Voice of Kenya, the British-controled national radio. The language that he knew so well would eventually find him as a translator when Kenyatta was tried by the British for advocating land reforms.

---

* There is no tribe called Mau-Mau; it was a derogatory term applied by whites, and it stuck.

Violence grew when Kenyatta was jailed by the British in 1952, and Leakey became a target for "Mau-Mau" terrorists; he wore a revolver and was accompanied by a bodyguard. (Nearly four decades later, in 1990, his son Richard, as director of the Kenya Wildlife Services, would be accompanied by armed bodyguards after he fired hundreds of corrupt government employees in the Kenya Game Department. See Chapter 9.)

When the Leakeys built a new house in the Nairobi suburb of Langata, the atmosphere of violence dictated its design. A center courtyard allowed the dogs to go out on their own at night, and heavy wire mesh covered the windows. Mary slept with a pistol beneath her pillow. The fervor of paranoia among whites was much greater than warranted; less than 40 were killed, a fraction of the number of Africans killed during the rebellion. Yet the threat to the Leakeys was real, and nearby Tanzania provided not only rich research sites but an escape to a more peaceful setting. For Mary Leakey, Tanzania was a world of beauty and discovery.

RETURN TO TANZANIA

While planning their return to Olduvai Gorge, Mary received a grant to study the rock paintings of the Kondoa district. Mary considers the time she spent in Kondoa as "one of the highlights of my life and work in East Africa." The woodlands and the granite boulders provided an idyllic setting, and the art itself was enchanting to study.

One of the reasons Mary was so delighted with this ancient art was that it depicted details of Stone Age life that rarely survive in the archaeological record. The paintings included bows and arrows, even hairstyles and jewelry. People danced, played musical instruments and hunted. The scenes were lively and offered a glimpse of how people lived—a far cry from excavating the cremated bodies at the Njoro River cave in Kenya.

By studying the details of the paintings, Mary learned something about the artists. She noticed that they were keen observers of wildlife; snakes were drawn larger than life, perhaps to denote danger; one artist conveyed the subtle difference in species of rhino. (The "black" rhino has a pointed lip, for browsing leaves;

the "white" rhino uses its wide lip for "mowing" grass. The feeding habits of animals are often revealed by physical details; this is why the study of hominid teeth is so important.)

To record the images, Mary traced the artwork on clear cellophane while perched on scaffolding built alongside the rock face. Back at camp during the evening, she transferred these drawings to half their scale on tinted paper. During the day she returned to the site to reproduce the exact colors.

The original paintings on the rock surfaces were often in a superimposed jumble, because many people had painted over the same rock canvas over the years. Mary had to find the images that belonged to the same scene or a single artist. Some might consider it tedious work, but Mary was delighted by the challenge. While she worked at tracing the figures, Louis and a group of friends set out on foot to find more sites, assisted by the local Warangi people. In the end, they found 182 sites, and Mary, assisted by Giuseppe della Giustina, traced 1,600 figures—all within three months' time.

By making small excavations around the area, she discovered "pencils" of ocher and other colorful pigments that had been ground down and mixed with grease. Other pencils found nearby were later dated at 29,000 years old. Because the paintings themselves were exposed to the air, the oldest that survived were probably only a few thousand years old.

A few reproductions were eventually mounted on rocklike contour surfaces for an exhibit at the Nairobi Museum, where they can be seen today. In 1983 Mary Leakey published a richly illustrated book entitled *Africa's Vanishing Art*. Her record may be all that remains; many of the original paintings have been destroyed by vandals or visitors who splash water on them to bring out the colors to take a photograph.

Ironically, the man with a keen interest in ancient art gave funds not for the Kondoa project but for Olduvai. In 1950 the Leakeys had visited France to meet with their supporter Charles Boise. As a token of gratitude, they took him on a tour of some of the rich archaeological sites and rock paintings, including the famous caves of Lascaux, as well as some of the sites that Mary had visited as a girl. They also invited Boise to visit them in Africa.

The following year, during his stay at Olduvai Gorge, Boise assured them that he would continue to support their work—for the next seven years. It was an incredible opportunity to do some long-term research rather than moving from site to site. The key to solving mysteries in human origins, or any science, is to focus on certain questions. At Olduvai Gorge, the Leakeys hoped to find "living floors," evidence of campsites, where bones of our ancestors were found alongside tools and the bones of animals that were left over from a meal.

Since their initial 1935 expedition, they had gone to Olduvai on short research stints, whenever they had the chance. But much of their work was limited to brief, surface excavations. Now the Kenya government supported the Miocene work on Rusinga Island. So with Boise's support, they could plan detailed, long-term research at Olduvai.

OLDUVAI

The Leakeys set up a tented camp under thorn trees in what's known as the Side Gorge, an angular branch of the major Main Gorge. Water remained a problem; they had to transport it from 30 miles away. Heselon Mukiri was in charge of the digging team, and Jonathan and Richard Leakey joined their parents during school vacations. One of the young assistants at Olduvai was Jane Goodall, who later became famous for her studies of chimpanzees in Tanzania.

The Leakeys focused on the level known as Bed II, the second oldest of five units at Olduvai, where they had already found several rich sites. At one of the sites they found over 2,000 tools and many mammal fossils. At a second, there were even more stone tools, and among the animal fossils, they found a curious collection of giant hooved animals with horns that spanned six feet across.

One of these animals had been found during the 1913 expedition led by Hans Reck, who named them *Pelorovis*, which means "monstrous sheep"; now these mammals are thought to be an ancestor of the buffalo. The curious thing about the Leakeys' discovery was the number of animals found together.

A cast of a *Pelorovis* skull at the Olduvai Gorge Museum in Tanzania. With horns spanning up to six feet, these extinct creatures are thought to be ancestral to the buffalo. (© Delta Willis)

Many bones of different individuals were found in a clay-filled gully; one large skeleton was discovered standing upright. Shattered leg bones were found alongside tools. Louis suggested that hominids might have driven animals into a bog, killing the smaller ones, while the largest ones became mired in the mud. "Many of them had been butchered," wrote Mary Leakey. "If their death in the swamp represented a hunting episode, it would have required bravery, skill and co-operative effort," because the animals were so large.

Other giants emerged from Bed II, including a pig as large as a hippo. Its tusks were three feet long; at first they thought they had found the remains of an elephant. They uncovered bones of a giant baboon, the size of a gorilla. They also found evidence of ancient horses or zebra as well as an extinct jackal, which inspired Louis to wonder if they had found the true ancestor of the dog. (Dogs and jackals are closely related; both are in the genus known as *Canis.*)

They also found the skull of a big cat, which Louis suggested may have been a tiger rather than a lion.

When he was a student at Cambridge, one of the laboratory assistants told him not to worry about looking at a big cat's teeth. "Just put the jaw on the table and try to rock it. If it rocks, it be a lion, if it don't it be a tiger." The jaw of a tiger has three points of contact with the surface of a table, while the jaw of a lion is curved, like the base of a rocking chair. The skull they found at Olduvai sat steady on their worktable.

Yet among all these animals and tools, they found no hominid fossils, until years later. In July of 1959, Heselon Mukiri was looking around in Bed I when he found a hominid tooth in a block of limestone. Once the matrix of limestone was removed carefully, they discovered that the tooth was attached to part of a lower jaw. Because there were many stone tools found nearby, they began to focus on Bed I.

It was important to the Leakeys to find hominid bones in association with tools. Louis felt that the use of tools was the hallmark of early humans; in fact, at the time, many scientists felt that only human beings made and used tools. Later Jane Goodall would observe chimps making and using simple tools.

When Heselon discovered the tooth and jaw, the Leakeys felt this new site in Bed I might hold more hominid discoveries. They delayed their search for a few days, waiting for the arrival of Des and Jen Bartlett, a documentary camera team, who were driving down from Nairobi with Richard Leakey, age 14. They were so confident about the potential of this site that they were sure the Bartletts would have something exciting to film.

Meanwhile, Louis became ill and stayed in camp to rest. A few years earlier he had suffered a heat stroke while working at Olduvai, and overnight, his hair turned white. So he rested, while Mary went off with her Dalmatians, to explore on her own. The date was July 17, 1959.

The site she explored was called FLK; Louis had named the gully, or korongo, after his first wife, Frida Leakey, before he had met Mary. Stones and bones had already been found in the area in previous visits; as she walked around, Mary noticed that the recent rains had uncovered even more.

Louis and Mary Leakey with the upper jaw of Zinj. Louis points to the teeth and palate. A modern human skull is in the foreground. (Des & Jen Bartlett © *National Geographic*)

One piece of bone caught her eye because it was not laying flat on the ground but was sticking out from beneath. She thought it looked like part of a skull, but the bones were very thick looking. She brushed away some of the deposit and was stunned to see parts of two large teeth in place in the upper jaw. Mary recognized the teeth immediately as those of a hominid. She was thrilled; she knew she had found a hominid skull, and she had a hunch that there was a lot of it there.

She rushed back to camp to tell Louis, who jumped out of bed and followed her back to the site. Mary thought the discovery was exciting, but Louis took one look and was disappointed. He recognized the skull as being like the australopithecines from South Africa. By this time Robert Broom and other researchers had found many australopithecines; some were delicate, like the Taung child, and others were robust, with dramatic features, like the skull that now lay before them.

This was not the "early man" that Louis wanted to find. He did not think that the australopithecines were a missing link; he described them as offshoots, not on the branch of the family tree that led to us. Mary had found a wonderful skull, but it was not the large-brained *Homo*. This skull had a small braincase. Mary recalls that Louis' first comment was "Oh, dear. I think it's an australopithecine."

The Leakeys covered the find with stones to protect it, so that it wouldn't be destroyed by a wild animal or the cattle of the Maasai. They waited for the arrival of the Bartletts. They filmed the excavation, which took 19 days. The Bartletts also took still photographs that would help publicize the find. This coverage would have a huge impact on the public, for around the world, people could grasp the excitement of a discovery.

A BOLD FACE

While Louis had been disappointed at first glance, he became more excited as the skull was uncovered. It was in very good condition and was not distorted, as *Proconsul* had been. And, as Mary had predicted, there was "a lot of it." As they excavated and sieved the area around it, they did indeed find a "living floor," with many stone tools and animal bones. Even cameraman Des Bartlett found a hominid leg bone.

Mary used some of the 400 skull pieces discovered to reconstruct many parts of the skull. The braincase was small, but it was marked by a ridge, or bony crest, at the top. This crest was evidence of a primate that had very strong and powerful jaw muscles, because it formed the anchor for those muscles; you can see an example of this in male gorillas. Gorillas spend most of their day eating leaves. This hominid spent a great deal of time chewing its food.

Louis Leakey gave it the scientific name of *Zinjanthropus boisei*. "Zinj" is the ancient Arabic word for East Africa; Zinjanthropus means "East African Man." The species name, *boisei*, was meant to honor their supporter, Charles Boise. Species names often come from people's last names or the area where discoveries are made. The name Louis chose would cause a controversy.

Why did Louis Leakey create a new name when his first impression was correct? At first glance, he had said he thought it was an australopithecine, and it was.

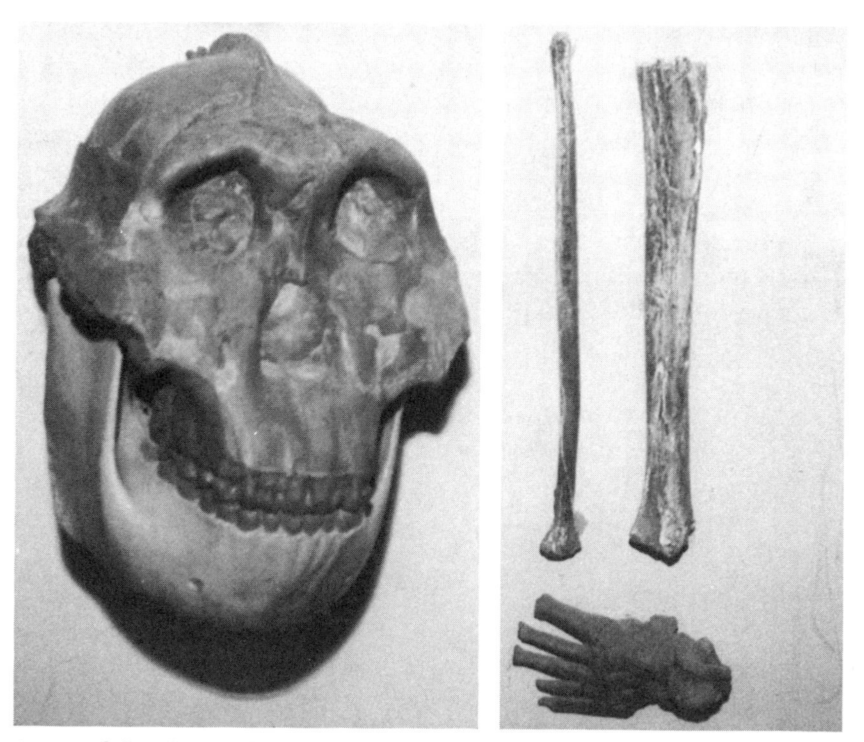

A cast of the Zinj skull appears alongside some of the remains of the Handy Man, or *Homo habilis,* at the Olduvai Gorge Museum. The discovery of the Handy Man forced Louis Leakey to change his ideas about Zinj as a toolmaker. (© Delta Willis)

However, Louis did not believe that *Australopithecus* made tools. Yet here was a very australopithecine-looking skull found right in the middle of plenty of stone tools. The tools—small pebble tools that had been given a sharp edge—were part of the older Oldowan culture. By giving the find a new name, Louis made it seem different from the australopithecines from South Africa.

In his first article for *National Geographic*, he described the teeth as "quite obviously human." The title of the article was "Finding the World's Earliest Man."

"Man—what does that word really mean?" Louis asked in the article. "To me it suggests no mere primeval apelike creature that walked erect and had hands. To be truly human, he must have had the power to reason and the ability to fashion crude tools to do his work. There is the key, the ability to make tools. . . ." Louis called the finds from South Africa "near-man."

The discovery was made in July of 1959; in August of that year, the Leakeys attended the Third Pan-African Congress, held in Zaire (then the Congo). They flew to the conference with the skull in a box resting on Mary's knees. The discovery created a great sensation among the delegates, who went to the Leakeys' room to examine the find in detail. But some scientists balked at the new name, especially those from South Africa, who recognized that this was an australopithecine. The name was eventually changed to *Australopithecus boisei*, and it is called *A. boisei* for short.

Louis had made a mistake by suggesting the new name, and this controversy, combined with his earlier claims about the Kanan jaw, did not help his scientific reputation. On the other hand, this skull was a much better and more complete specimen than any of the skulls found thus far in South Africa, and his colleagues were impressed by that as well as the extraordinary "living floor" that the Leakeys had discovered. The colorful record made by the Bartletts brought this new science of paleoanthropology to the readers of *National Geographic*, and the National Geographic Society began to fund the Leakeys research "on a scale that exceeded our wildest dreams," Mary wrote. The Leakeys received over $20,000 for exclusive rights to the magazine story, and that was only the beginning. National Geographic would support the work at Olduvai and Laetoli with grants over the next three decades. "Zinj" changed the Leakeys' lives and brought them more fame and fortune than they had ever known. The image of the skull appeared on Tanzanian postage stamps.

The skull also did something else. It moved the "cradle of mankind" from South Africa to East Africa. When the skull was first discovered, Louis thought that it was just over a half million years old. No one argued with Louis Leakey's claim that "the Olduvai skull represents the oldest well-established toolmaker ever found anywhere."

There was some confusion about the dates on the fossil finds from South Africa, and those fossils were not clearly associated with tools, as were the Leakeys' finds. Louis Leakey's claims for the "oldest" hominid were about to be proved, beyond *his* wildest dreams.

THE FIRST GOOD DATE

A team of geologists at the University of California, Berkeley, had been working on ways to refine the method of dating volcanic rocks. As mentioned, fossils this old cannot be dated directly, so geologists estimate their age based on the dates they can get from volcanic material in the levels above and below them. To do this, they take samples of the levels at the site. Pumice, the volcanic stone used for buffing heels and toes, is an ideal sample.

In the 1950s, Jack Evernden and Garniss Curtis of Berkeley developed a new method of dating. It is called the potassium/argon method, because it compares the ratio of these two elements. As potassium decays, argon accumulates. Potassium exists in many places; it's even found in the human body. And it decays all the time. But when a volcano is born, the potassium is brand new, and pure. From the time of a volcanic eruption, the potassium decays, and argon builds up. The more argon you have, the older the volcanic rock. The less argon, the younger.

When "Zinj" was discovered, it offered a good test for the Berkeley geologists. They went to Olduvai, took samples and returned to their lab to analyze the samples. The age of the skull was fixed at 1.75 million years old.

This was great news, for two reasons. Its age proved that the oldest hominids came from East Africa. This is what Louis had always believed, even though his professors at Cambridge advised him to look anywhere but in his own backyard. In additon, the new method could be applied to many other hominid skulls, or their sites. Remarkably, many skulls have since been misdated, because samples can be impure, and people can make mistakes in labs. But the date for "Zinj" has remained solid throughout all these years. Twently years later, Garniss Curtis would be involved in solving one of the most controversial dates ever, on a skull from Kenya found by Richard Leakey's team. (See Chapter 8.)

Today the Zinj skull is at the Tanzania National Museum in Dar Es Salaam. Its discovery made the Leakeys famous, and the money from the National Geographic Society gave them a chance to find even more hominids at Olduvai.

But after the discovery of Zinj, Louis became involved with other projects, including fund raising. Beginning in 1960, he went on a long lecture tour in the United States, which he would do annually for a dozen years thereafter. He began to develop a primate center, and inspired the work of primatologists such as Jane Goodall and Dian Fossey, who studied mountain gorillas. He continued to develop the Nairobi Museum, creating an institute for the study of paleontology and archaeology. And he returned to work on the Miocene sites in Kenya.

Mary Leakey continued to work at Olduvai. Richard helped her build a permanent camp on the north side of the Main Gorge, just across from the Zinj site, or FLK. Mary lived there most of the time, making only short visits to Nairobi. Louis visited on weekends when he could, and Mary missed him very much.

Mary had plenty of work to do, exposing 3,600 feet of the "living floor" around where Zinj had been found. She was assisted by a team of 16 African workers as well as visiting scientists. In the summer of 1960, her oldest son, Jonathan, helped section off an area of FLK into grids. Jonathan watched a team uncovering some bones nearby. After studying one, he asked his mother, "Does any animal have a long thin bone like this?" and traced the shape he was looking at with his finger. Mary replied that she couldn't think of one. "Oh, then I think it must be hominid," the 20-year-old commented.

Mary dropped her drawing and rushed to see. Jonathan pointed to a hominid leg bone. Thus began an excavation in an area called "Jonny's site." The leg bone that he found led to the discovery of foot bones, including a toe, as well as a jaw. Slowly a new kind of hominid emerged, very different from Zinj. The skull bones were much thinner, and the braincase was larger. What would they name this new species?

CHAPTER 6 NOTES

p. 60 "one of the highlights . . . " Mary Leakey, *Disclosing the Past,* p. 105.

p. 63 "Many of them had been butchered. . ." Mary Leakey, Disclosing the Past, p. 119.

p. 64 "Just put the jaw on the table. . ." Louis Leakey, *Animals of East Africa*, p. 163.

p. 66 "Oh, dear . . ." Mary Leakey to Roger Lewin, *Bones of Contention*, p. 138.

p. 67 "quite obviously human . . ." Louis Leakey, "Finding the World's Earliest Man," *National Geographic*, September 1960, p. 421.

p. 68 "on a scale that exceeded our wildest dreams . . ." Mary Leakey, *Disclosing the Past*, p. 122.

p. 68 "the Olduvai skull represents . . ." Louis Leakey, as quoted in *The Times* (London), September 4, 1959.

p. 70 "Does any animal have a long thin . . ." Mary Leakey, *Disclosing the Past*, p. 126.

7

THE HANDY MAN

The new discovery from Olduvai was the first evidence of the large-brained toolmaker that Louis Leakey had always hoped to find. It had the added attraction of being 250,000 years older than Zinj. Some of the hand bones suggested a precision grip, an ability to grasp tools with an opposable thumb. The skull itself was more human looking, without the crest, without the apish brow ridges. The teeth were smaller, more human. Placed alongside a modern human skull, it was a lot easier to imagine this as one of our ancestors than Zinj.

Louis promptly dropped his earlier claims about Zinj; suddenly it was no longer the "earliest man" but, like the rest of the australopithecines from South Africa, a "near-man." Now the real toolmaker had been found. The name Louis Leakey, Phillip Tobias and John Napier chose was *Homo habilis*, which means "Handy Man." By the time this new name was announced in 1964, two other skulls had been found at Olduvai to strengthen their evidence.

Yet some scientists argued that these older bones from Olduvai were just an advanced form of australopithecines—despite the larger brain. *Homo*, which means "man" (or "humanity") in Latin, is the genus associated with large brains and increasingly intelligent primates. Our own subspecies name is *Homo sapien sapiens.* Almost everyone agrees that the steps in evolution for our kind went from *Homo habilis* to *Homo erectus*, then to *Homo sapiens.* But the naming of the Handy Man created even more controversy than the naming of Zinj.

The new species challenged the theory that humans descended from the australopithecines. Louis never believed that theory anyway. He always believed that *Homo* would be found far back in time, so this new discovery from Olduvai, nearly 2 million years old, pleased him.

Ironically, it would be Richard Leakey's team that would find evidence that suggested his father's ideas were right.

A YOUNG REBEL

Richard challenged his father as soon as he was capable of speech, and eventually this led to competitive, political maneuvers. Mary noticed that Louis was less patient and less caring toward Richard than his other sons (now including Philip Leakey, born in 1949). But they had much in common, and Richard's adult life patterns would be remarkably similar to his father's.

Both separated from their first wives shortly after the birth of a child, to go on field expeditions with the women who were to become their second wives and real partners. Both were stubborn and stoic when confronting serious health problems. Both had an extraordinary appeal to the public—especially women and students—and a talent for raising money. They shared a love for Africa and Africans.

In Kenya, Richard learned to speak Swahili fluently before he studied English, and developed close friendships with many Africans. On his first day in junior high school, he was tormented by the other students, all white, for his friendships. His classmates locked him inside a wire cage, about three cubic feet.

> *I was crouched like a monkey in this tiny cage, with no way of escape. . . . I was poked with sticks, spat upon and even urinated upon, for what seemed an eternity until it was time for everyone to go to assembly. I remained in my cage, very miserable and frightened. . . . I was eventually released by a senior master who had to get a hacksaw to cut through the padlock. He had no doubt that I was to blame; so, wet through, filthy and stinking, I began my first day of senior school.*

Richard displayed the same enthusiasm for the formal classroom as his mother had two decades earlier, and although he eventually qualified to attend college in England, he left when he became homesick for Africa.

He loved going on safaris and, like his parents, developed a passion for wildlife and wild places. As a boy he trapped animals and captured snakes; a bite from a puff adder almost proved fatal. It wasn't the poisonous snake bite, but after a doctor followed up the antivenom with a shot of horse serum, Richard woke in the middle of the night to find that he was totally paralyzed; he couldn't get out of bed, nor could he call for help. Years later he found out that he was allergic to the serum and had nerve damage that make his hands shake.

He had better luck capturing lions, which he did by putting bait in cages with trapdoors, and found that he could earn money collecting animals for filmmakers like the Bartletts. He began to collect small primates, including pottos and bushbabies, for zoos in Europe and the United States. His capture of baboons relied on imitating the sound of a leopard; he would slip into the woods before dawn, and his leopard growl sent baboons running out of their roost, into his reach.

At the age of 17, Richard was not sure of what he wanted to do for a career, but he felt that he didn't want to follow in his parents' footsteps. "I was determined to distance myself from my parents and their work on fossils and prehistory, largely because I wanted to be my own man."

In 1960 and 1961, when a long drought killed huge numbers of animals in Kenya and Tanzania, Richard saw an opportunity to sell skeletons to universities and museums. He borrowed $1,000 to buy an old Land Rover and drove around collecting carcasses, which he boiled down in vast steel drums. Because he had to label the bones, he learned a great deal about anatomy, a skill that he would find useful later.

In 1962 he worked as a camp manager at Olduvai, organizing supplies. When officials of the National Geographic Society visited, Richard served as their safari guide. They, in turn, recommended him to many friends who wanted to go on safari, and for a few years he ran a tour company for photographic safaris with filmmakers Alan and Joan Root. (The Swahili word *safari* simply means a "journey of adventure," not necessarily a hunting safari.

Richard was not a "white hunter," leading safaris to shoot game for trophy.)

One of Richard's jobs was to run a camp for the National Geographic film team at Olduvai. The filmmakers had a small airplane for shooting low aerial shots, and Richard was invited to go aloft. He was terrified of flying; the pilot knew this and made a series of abrupt banks and sharp climbs around the gorge that gave Richard white knuckles. Once he was safely on the ground, he decided that the only way to overcome his fear of flying was to become a pilot. The following year, in 1963, he made his first solo flight.

THE FIRST EXPEDITION

That same year, on a flight from Nairobi to Olduvai, Richard noticed the exposures, or eroded areas, around Lake Natron, which his mother had found a few years earlier returning from Olduvai by car. With his parents' encouragement, Richard set out on his first official fossil survey. "It was just an adventure at that stage,"

Twenty-one-year-old Richard Leakey accompanies his parents to the National Geographic Society in Washington, D.C. Richard shows the Peninj jaw to Dr. Melvin M. Payne, right. (Winfield Parks © *National Geographic*)

Richard explains. He liked planning the expeditions, and camping out, more than studying bones.

With archaeologist Glynn Isaac, Richard spent two weeks exploring the western shores of Natron, which is just below the Kenya/Tanzania border. They called the fossil-rich area Peninj, after a river that flowed nearby. With financial support from his father's National Geographic grant, they returned in 1964 with a small team of African fossil finders. One of them, a member of the Kamba tribe, was named Kamoya Kimeu. Kamoya became one of Richard's best friends, and they would work together in the field for 30 years. Like Heselon Mukiri, Kamoya has a keen eye for hominid fossils.

That year, Kamoya found a complete lower jaw of an *A. boisei.* At the time, it was the only jaw known for the same species as the famous Zinj skull that Mary Leakey had found at Olduvai. The Leakeys were delighted by the discovery, and plans were made for a three-month survey, to pinpoint other fossil sites around Lake Natron. The discovery of the Peninj jaw gave Richard a new view. "I discovered that the credit for the research, the scientific publications, was going to people who I didn't consider had done nearly as much work as I had, setting up the expedition. But I wasn't a scientist; I was just a young kid, fooling around."

Richard began to like the scientific expeditions better than dealing with tourists. He also liked one of his mother's assistants, Margaret Cropper, who had accompanied the team on the 1964 expedition to Lake Natron. In 1966 Richard and Margaret got married and began to work together excavating fossils near Kenya's Lake Baringo.

Margaret was trained as an archaeologist. "She was the brains of the project, the scientist," Richard recalls, "and I was the provider," running the camp. With Margaret's academic background, the two co-authored a scientific paper on a new species of an extinct monkey. They had unearthed a skull and nearly complete skeleton of a primate related to the modern-day colobus, the beautiful long-tailed monkey with a black-and-white coat.

By this time, Richard had become interested in primate fossils and had read many technical papers. "Having read these papers, I couldn't understand for the life of me why you had to have a degree to write a descriptive paper about a set of new fossils. It struck me

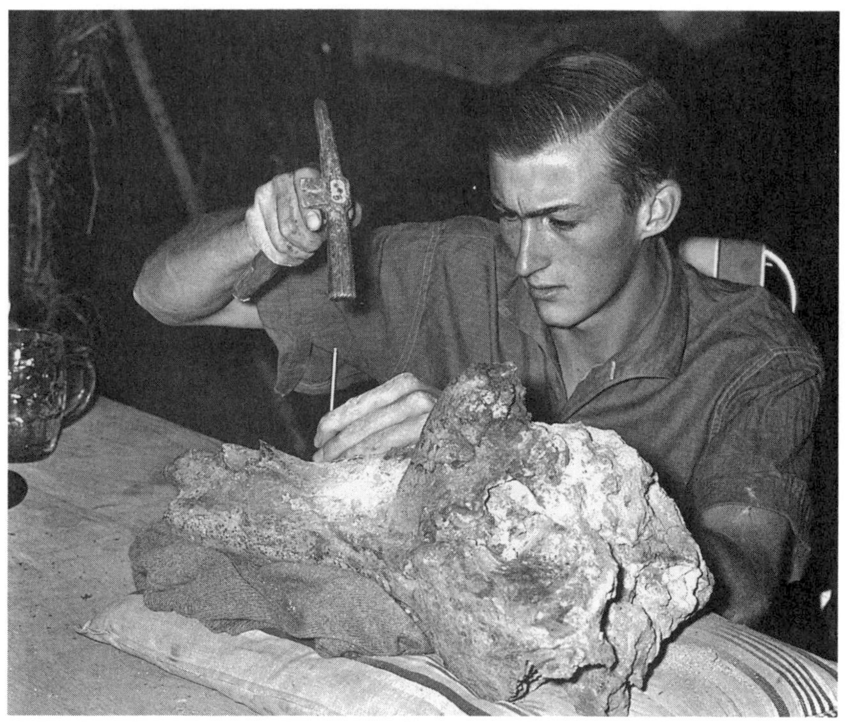

Richard Leakey removes the matrix from a pig fossil discovered at the Peninj site at Lake Natron, in Tanzania. (Bob Campbell)

as very odd, because it was simply looking and reading and writing and using anatomical terms, which were, first of all, not foreign to me because I'd grown up amongst them, and secondly, were available to me from any textbook on anatomy." The paper was published in 1967, and it was well received. For Richard, "It sort of broke through the mystique."

"I'M THE DIRECTOR"

The paper was Richard's first real step toward acceptance in the scientific field. His next step was even more bold. "I made a definite decision that if I was to make a career of paleontology, without a university degree, I needed to have an institutional base of my own." He had helped his father at the Center for Prehistory and Paleontology, which is part of the Nairobi Museum, and even established a fund-raising society to help finance projects. In 1968 he began to lobby for the post of director of the museum. Since Richard was Kenya-born and had already showed innovation as a part of the administration, he got the job. He was 23 years old.

His new power brought him into conflicts with his father. It was Louis' policy that all important fossils be sent to England for casting, or reproduction. Richard wanted to have local Kenyans learn this skill. Part of his plan for the museum, as well as all field research, was to train Kenyans, rather than giving the work to foreigners. "I suggested we bring in some English technicians and teach our own people how to do it." But Louis had already made plans to send the finds to colleagues in England. "He'd made some commitments and tried to push me." In the end, Richard said to his father, "Look, I'm the director." It became a debate that would be solved by a government minister. Richard says simply, "I persuaded the minister; he didn't," which is another way of saying "I won."

There had been other conflicts. When his father was away for several months, Richard, left in charge of the Center for Prehistory and Paleontology, ran things the way he wanted to, changing his father's policy. As director of the museum and leader of many research projects, he dismissed many of his father's colleagues, despite their seniority. He wanted to have his own team. At the age of 24, he would.

THE OMO EXPEDITION

The Omo River Valley is just north of the Kenya/Ethiopia border. The muddy river flows for 700 miles, from near Addis Ababa in the north, down into Kenya's Lake Turkana. The lower valley is a rich delta. These sediments buried bones, creating thousands of fossils, and the changing course of the river exposed them. The French were the first to explore the ancient riverbanks; during the 1930s, an expedition led by Professor Camille Arambourg collected over 100 tons of fossil in two years.

During World War II, Louis Leakey took advantage of the protection of British troops along the lower Omo and sent his most trusted assistant, Heselon Mukiri, to look for fossils. In 1959 American paleoanthropologist F. Clark Howell also explored the region. Howell and Leakey began to plan an international expedition to the area. But by the time they gained permission from the Ethiopian government to do so, Louis had arthritis in his hips, which made fieldwork difficult. He appointed his son Richard as the field leader of the Kenya team.

Howell led the American team, and Professor Arambourg led the French team. While they didn't find many good hominid fossils, the Omo expedition was a turning point in important ways. It brought scientists from many disciplines into the field to work together in a massive international effort, making the "field a laboratory," as Howell explained. Not only were there geologists on the team, but paleontologists in search of baboon fossils, and even scientists who collected fossil pollen, to determine what kind of trees and vegetation lived in the area millions of years ago. Many of these scientists, like Leakey, were young; most were graduate students. But as the leader of the Kenya team, Richard Leakey found himself frustrated.

"I didn't like the Omo project," he recalled years later. "There were too many chaps ahead of me—all sorts of serious, senior scientists. I was very much on the bottom of the pile." He wanted control of his own team and his own fossil expedition. The way he achieved his goals was fast and forceful, the same way he became director of the museum. "I needed a project that was good, big and attractive."

During the Omo expedition, he flew south to Nairobi to collect more supplies. On the way back, the pilot of the chartered plane detoured around a storm, flying over the eastern shores of Lake Turkana. Richard looked out the window to see vast stretches of exposures—light-brown sediments cut by erosion—the kind of exposures that reveal fossils.

A few days later he borrowed the helicopter from the Omo team and returned to East Turkana. Within minutes of landing, he held a stone tool. Nearby he found a fossil jaw of a pig. He decided to explore a few more places, telling the copter pilot to touch down here, touch down there. "Everywhere we stopped, there were fossils." In his excitement, he forgot to note the exact location

Richard and Louis Leakey discussing fossils found during the 1968 Omo expedition to Ethiopia. The following year Richard left the Omo to make his own discoveries near Lake Turkana in Kenya. (Bob Campbell)

where he had found the jaw and the tool, or any of the other sites. He knew only that he was somewhere northeast of the lake.

A few months later Richard accompanied his father to Washington, D.C., to make a report on the Omo expedition to National Geographic Society trustees. Without warning Louis beforehand, he boldly suggested to the trustees that the $25,000 that was to go to the Kenya team in Ethiopia be given instead to his own project at Lake Turkana. Surprised by his son's proposal, Louis spoke out against it, arguing that there were plenty of fossils to be found on the Omo, and Richard was just young, in search of instant prizes.

The trustees excused the two from their boardroom. Behind closed doors, they admired Richard's "cheek and initiative." Richard was awarded the grant. Richard's new plans for East Turkana took the money away from Louis' plans for the Omo.

"Richard was always a competitive person," Mary notes. "When he entered a new field it was with the intention of getting to the top, and the sooner the better. And who at the time was in possession of the summit, needing therefore in due course to be replaced? Louis."

Louis was also beginning to feel envious of Mary's achievements at Olduvai Gorge. She did most of the work, and scientific colleagues were beginning to recognize her contributions and how her approach to science was much more careful than Louis'.

Mary made no bold claims about what discoveries might mean, but always worded her views in cautious language. Louis did just the opposite at a controversial site in California, called the Calico dig. It wasn't the first time Louis and Mary had disagreed on scientific questions, but before they had always debated in a friendly way. Now Louis was being so careless and so stubborn that Mary lost her professional respect for him. Their marriage was already strained because they spent so much time apart; now their professional partnership began to fall apart.

"I became the opposite of a comfort to him; a cause of exasperation, a harm to his self-esteem," Mary wrote years later. Consequently, Louis sought approval from the audiences in California, who were thrilled to have their own "early man" site. At the same time, he pushed himself physically, and his health went downhill fast. In addition to his painful arthritis, he had high blood pressure and, in 1970, had his first heart attack. He never allowed himself any rest and ignored his family's warnings to "take it easy."

"He still had tremendous ambition and energy," Richard recalls, "and I think he found us to be weights around his neck, saying you can't do this or that—trying to help him survive. In California, everyone adored him. There he got the hero worship that he preferred to us saying 'no' to everything. I can understand it."

At the nadir of his career, Louis enjoyed the public acclaim of a scientific hero. Some of his bold and outrageous hunches had, after all, turned out to be right. He had insisted on searching in Africa, when other scientists said that missing links were not to be found there. He had inspired several young women (with no prior academic training) to go into the field and study primates. People made jokes about "Leakey's Ape Women," yet Jane Goodall and Dian Fossey would become famous for their work and contribute articles to *National Geographic*. Both women eventually obtained academic credentials; Dr. Goodall recently completed 30 years of field studies of the chimpanzees of Gombe National Park in Tanzania, the longest field record of any primatologist. Dian Fossey, who was portrayed in the movie *Gorillas in the Mist*, is remembered for her stance against poachers who were killing the endangered mountain gorillas in Rwanda and Zaire.

To most of his admirers, Louis was simply taking another bold chance in California. His critics were not only in the minority, they were concentrated around his own home in Kenya, where Richard had moved far from "the bottom of the pile" and into the spotlight. In his autobiography, he entitled this chapter of his life "Out of My Father's Shadow."

KOOBI FORA

Koobi Fora is the local name for a long, sandy peninsula that curls out into Lake Turkana. It is located on the eastern shore, 10 miles above Allia Bay. With the cool breezes that travel across the lake and the herds of zebra and antelope that come to drink, it provides a welcome retreat from the volcanic badlands that stretch for hundreds of miles to the north and the east. This harsh, arid terrain is swept by strong, hot winds, and around 2 o'clock in the afternoon, temperatures can reach 105° in the shade. The search for fossils usually begins at dawn, with a break in the middle of the day for a nap, and resumes again in midafternoon until dusk.

Richard eventually cleared a dirt airstrip at Koobi Fora and built a permanent camp of thatched-roofed, flagstone huts, or *bandas*.

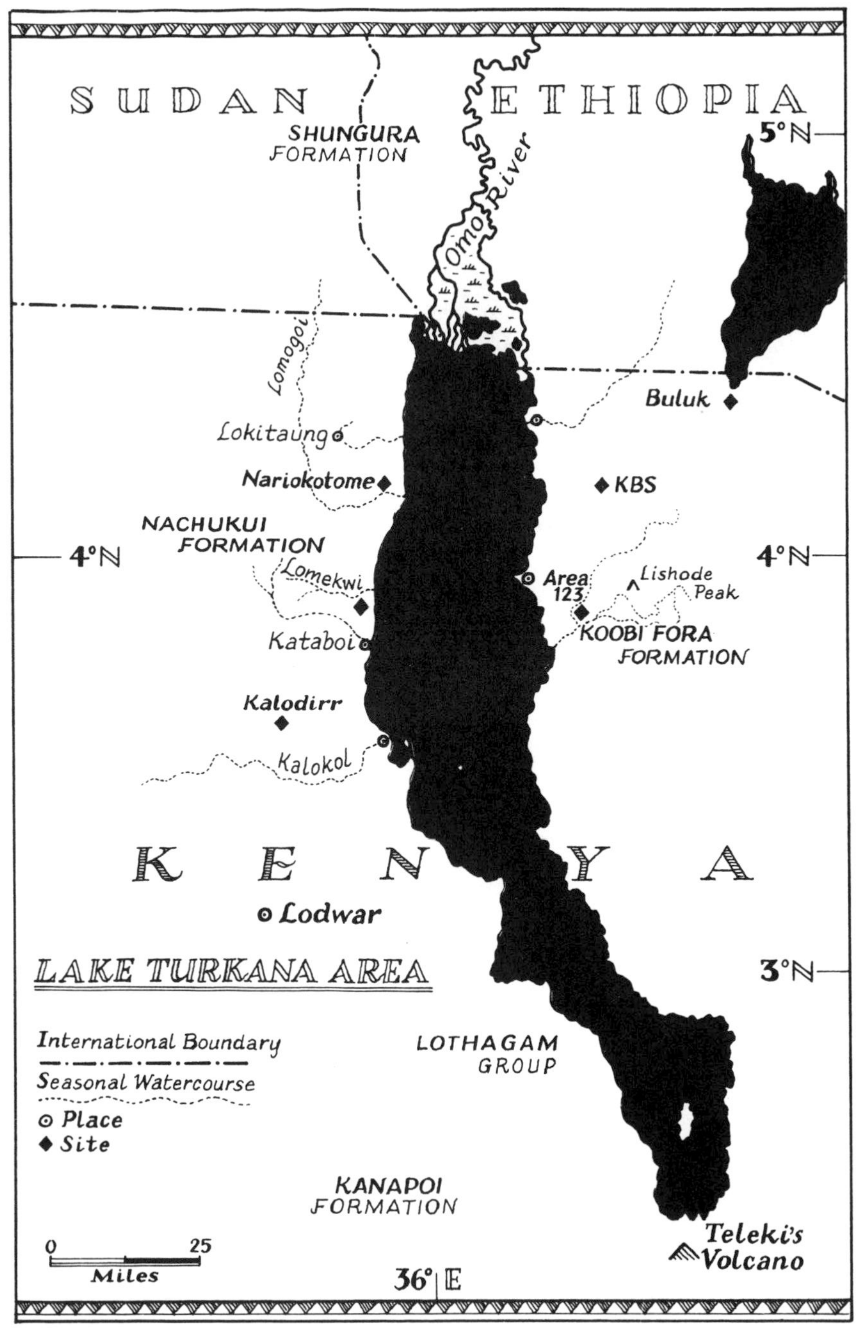

Map of Lake Turkana Area (Bob Gale, © Delta Willis)

The Koobi Fora base camp on the edge of Lake Turkana. The thatched-roofed, flagstone buildings feature large windows to admit the breezes from Lake Turkana. The large building on the right includes a dining room in the center and a laboratory on the left. (© Delta Willis)

The eight *bandas* have two sleeping cots each and long windows on both sides. There is a large central hut, with a dining table, and a simple laboratory, where bones are kept on shelves and in boxes before being transported back to the Nairobi Museum in Richard's plane. Meals are prepared over an outdoor wooden fire, and water is transported from some distance away; Lake Turkana waters are too rich in algae and alkaline for drinking, but supply plenty of fresh fish.

During the first 1968 expedition, Richard's team found enough hominid fossils to justify continued support. His 1969 visit to Washington, D.C. included a report of his success to the National Geographic Society as well as his first public lecture, to an audience of 3,000 people, which Richard found "exhilarating."

Over the next decade, the 500-square-mile region of East Turkana was surveyed by a number of scientists as well as the African Hominid Gang, led by Kamoya Kimeu. Working for six months of

Meave Leakey helps Richard reconstruct a fossil at the Koobi Fora camp. Their daughter Louise watches. (Bob Campbell)

the year, during the dry seasons, they recovered hundreds of fossils that had been eroded by wind and rain. Kay Behrensmeyer, a geologist from Harvard, joined the team in 1969, as did Meave

Epps, a British paleontologist who would marry Richard after his divorce from his first wife, Margaret.

Satellite camps were often set up in the field, closer to major fossil sites. As mentioned in Chapter 1, camels were initially used for surveys. During one camel-back survey in 1969, Richard and Meave Leakey discovered the remarkably complete *A. boisei* skull mentioned earlier. Behrensmeyer found some flake tools while doing a geological survey, and the site was named after her, KBS. The KBS site was to become the center of a huge controversy when an extraordinary skull was found nearby.

SKULL 1470

Unlike the 1969 discovery, the skull that came to be known as 1470 would emerge in over 300 pieces. The first clues were found on August 27, 1972 by Bernard Ngeneo, a member of the Hominid Gang. They were light-colored fragments, none more than an inch long. Some fragments were from the top of the skull, some from the back and some from the sides. This collection, plus a few fragile facial bones, inspired a thorough sieving of the topsoil in the area, at the bottom of an eroded gully. More fragments were collected, wrapped in toilet tissue, then taken back to the Koobi Fora camp, where they were washed and laid out to dry.

The very first afternoon the team at Koobi Fora began to reconstruct the skull, they saw that it was larger-brained than the australopithecines found at Koobi Fora or anywhere else. Richard Leakey and other scientists on his team thought that they had found a superb example of the large-brained *Homo habilis*, or Handy Man. But for the moment, they simply called it skull 1470, its code number for filing in the museum archives. It was the 1,470th fossil collected at East Turkana.

Because geologists had dated the KBS site at 2.6 million years ago, and skull 1470 was found at a level below the volcanic tuff of that age, it was initially thought to be around 3 million years old. Over the next few years, debates about the age of the skull would cause a split in the search for human origins pitting Richard Leakey's Koobi Fora team against Clark Howell's Omo team. But for the moment, the discovery brought Richard and Louis Leakey together.

Richard flew skull 1470 down to Nairobi to show to his father. No discovery from Olduvai had been so complete, or so old, and Louis felt that here was final proof of what he had been predicting all along. The age of 1470 confirmed Louis Leakey's theories—that the large-brained *Homo* went far back in time. It was the skull that he had longed to find for 35 years. "It's marvelous," he told his son, "but they won't believe you!"

"He was delighted," Richard recalls. "I think that's one of the days I remember most about him—his absolute joy." The Leakey family spent the evening together at their Langata home. "It was almost like old times," Mary recalled; Louis was happy and extremely proud. "In many ways I felt closer to him that day than I had since my early childhood," Richard later wrote. "I had the distinct feeling that we had finally made a real peace."

That evening Richard drove his father to Nairobi airport; Louis was scheduled to begin another lecture tour in California and stop in London on the way. As they left Langata, Mary had the feeling she would never see her husband again. At the airport, Richard wished they had longer to talk. A few days later, on October 1, 1972, Louis Leakey died of a massive heart attack in London, at the home of his close friend, Mrs. Vanne Goodall, Jane's mother.

Mary had already returned to Olduvai and Richard to Koobi Fora. They received the news and returned to Nairobi, to gather for his funeral; less than a week earlier, they had been a family together in celebration.

CHAPTER 7 NOTES

p. 73 "I was crouched like a monkey . . ." Richard Leakey, *One Life*, p. 35.

p. 74 "I was determined to distance myself . . ." Richard Leakey, *One Life*, p. 54.

p. 75 "It was just an adventure at that stage . . ." Richard Leakey, interview with the author, 1982.

p. 76 "She was the brains of the project . . ." Richard Leakey, interview with the author, 1982.

p. 76 "Having read these papers . . ." Richard Leakey, interview with the author, 1982.

p. 77 "I made a definite decision . . ." Richard Leakey, interview with the author, 1982.

p. 78 "I suggested we bring in some English technicians . . ." Richard Leakey, interview with the author, 1982.

p. 79 "field a laboratory . . ." F. Clark Howell as quoted in Delta Willis, *The Hominid Gang*, p. 39.

p. 79 "I didn't like the Omo project . . ." Richard Leakey, interview with the author, 1982.

p. 79 "Everywhere we stopped . . ." Richard Leakey, interview with the author, 1982.

p. 81 "Richard was always a competitive person . . ." Mary Leakey, *Disclosing the Past*, p. 141.

p. 81 "He still had tremendous ambition . . ." Richard Leakey, interview with the author, 1982.

p. 84 "exhilarating," Richard Leakey, *One Life*, p. 110.

p. 87 "It's marvelous," Louis Leakey as quoted in Roger Lewin, *Bones of Contention*, p. 128.

p. 87 "He was delighted . . ." Richard Leakey, as quoted in Delta Willis, *The Hominid Gang*, p. 43.

p. 87 "It was almost like old times . . ." Mary Leakey, *Disclosing the Past*, p. 159.

p. 87 "In many ways . . ." Richard Leakey, *One Life*, p. 150.

8
CONTROVERSY

Debates are an essential part of science, because it's important to test ideas and the facts behind theories. To make certain that the facts are real, scientists are supposed to share their data, offering their fossils to other scientists for review and writing their reports in such detail that any other scientist can test their facts and figures.

For example, when a geologist like Kay Behrensmeyer writes up a report on the layers of volcanic ash at the KBS site, she gives enough directions as to their exact location so another geologist can follow in her footsteps and test her research. This is why scientific papers are so technical. The laboratory results of the potassium/argon numbers have to be printed in detail, so that other lab tests can be done independently.

Louis Leakey was involved in many debates, and certainly his ideas were challenged. The devastating criticisms by Percy Boswell on the Kanam jaw might have crushed a person with less confidence. Raymond Dart's claims on the Taung child were criticized so severely that Dart withdrew from the debate circuit and stopped looking for more australopithecines. "It's no good being in front if you're going to be lonely," he said. Scientists have feelings just like anyone else, and sometimes their emotions get in the way of their work.

After Louis Leakey's death in 1972, Richard and Mary Leakey were both confronted with bitter and intense attacks. Some of the criticisms had nothing to do with their work; they were little more than hateful gossip meant to disparage their reputations. This often happens when people are famous, and the attacks usually come from people who are jealous. Two American paleoanthropologists devoted a great deal of energy to criticizing and challenging what they called "the Leakey dynasty." Most reporters and TV producers love this sort of rivalry, and emphasized the conflict of personalities rather than the scientific import of what had been found or what was really known about these discoveries.

The end result was that both Richard and Mary Leakey withdrew from public debates. In 1979, Richard became very ill and nearly died. He vowed that if he lived, he would leave the field of paleoanthropology.

A SECOND LIFE

In September of 1968, just a few days before Richard was to begin his job at the Nairobi Museum, he developed a sore throat and fever. His doctor discovered that he had a kidney disease and suggested he rest for six weeks. Richard, afraid that the job offer might be withdrawn if he were absent for so long a time, was determined to begin as scheduled, so he ignored his doctor's advice and went to work.

His kidneys were permanently damaged, and it was just a matter of time until they failed to work at all. At some point he would have to have a transplant or move to a place where he could undergo regular dialysis—which meant leaving Africa.

Richard pushed it out of his mind and decided to try to live a normal life. He didn't tell anyone about his illness except his wife. At the time, this was Margaret, who was shocked that he would abuse his health in favor of his career. Margaret didn't appreciate the intensity of his ambition; according to Richard, she didn't think he should have been offered the job at the museum; he was much too young and inexperienced. That may have been true, but above all else, Richard valued loyalty, and like his father fleeing to California, he found loyalty elsewhere—in his second wife, Meave, in fossil finders such as Kamoya Kimeu and in his Koobi Fora scientific team.

For the next 10 years, he worked harder than ever, even though he had severe headaches and nausea. He had dark circles under his eyes and was thin; he didn't like to eat because of a bad taste in his mouth. In 1979, following a public debate in Philadelphia that emphasized his "rivalry" with American paleoanthropologist Donald Johanson, Leakey became extremely ill, and eventually went to a London hospital, to begin the painful process of dialysis. His brother Philip offered to donate one of his kidneys, and the transplant was successful, or so it seemed.

On Christmas Eve of 1979, three weeks after the operation, Richard's body tried to reject the kidney. He was given drugs to suppress his immune system, which created a situation something like AIDS, because his natural defense system was put out of commission to allow the foreign organ to stay. This made Richard vulnerable to any and every disease; he promptly caught pneumonia, and his condition was so serious that one newspaper referred to "the late Richard Leakey." Meave gave him the strength and hope to live, and after his return to Kenya, other scientists gave him the confidence to go back into the field. He began what he called his second life.

While he was in the hospital in London, there was an attempt by a detractor to take away his job at the museum, which Richard had built into the largest institution of its kind in Africa. When he first began, there were 27 employees, but in 1979, there were around 300. He had increased the annual budget from $50,000 to over $1 million, with "considerable effort," raising many of the funds from foreign donors and foundations. Beyond the growing complex in Nairobi, annex museums and educational centers had been developed in Lamu, Kitale, Mombasa and at archaeological sites such as Olorgesailie.

Richard was bitter about the attempts to displace him, especially because "the organization I had built was to be broken up and I was not in a position to fight back." He found a way to fight back and spent a small fortune on long-distance telephone calls. The conflict was resolved, but there were several moments when he must have felt trapped, as he had been in that cage at school.

At the 1979 Philadelphia meeting, he had felt trapped by a reporter from the *New York Times* who forced him to answer a question about Donald Johanson's discovery, "Lucy." Leakey disagreed with Johanson's ideas about the 1974 finds from Ethiopia, but he felt the issues were better addressed in scientific papers rather than popular forums.

The Leakeys' foremost critic, Johanson had begun what was described as the "first real and successful challenge to the 'Leakey dynasty' ever made." Not only did he believe that Lucy (an older australopithecine) disproved the Leakeys' ideas about our recent ancestors, but Johanson was especially critical of Richard Leakey's role in a heated controversy about the KBS site, where skull 1470 was discovered. The skull was initially estimated to be 3 million years old, then revised to be 2 million years old.

THE KBS CONTROVERSY

Despite the early success of dating Zinj at Olduvai, the methods for dating fossils are still being refined. The geology of many hominid sites is complex and confusing, because most of them are in the Great Rift Valley.

The Great Rift Valley stretches through seven countries, from the southernmost tip of Turkey to Mozambique. It is a huge ragged scar in the face of the earth, visible from the moon. The rift began around 14 million years ago when the earth's crust began to pull apart from east to west, a result of the shifting tectonics (or the movement of geological plates) that shaped our continents and continue to shape them today. The rift is still pulling apart—only a few inches a year—but if it continues, millions of years from now the Red Sea will flow down into the basins and salt water will cover Olduvai, Naivasha, Nakuru, Baringo and the basins around Lake Turkana.

The dramatic movement of the rift can be traced in fault lines, such as the San Andreas fault of California, or the red ones you can actually see at Olorgesailie. These fault lines mark places where the earth's plates shift and collide, and their collision often leads to earthquakes and volcanoes. The greatest volcanic activity in the African Rift occurred in what is known today as Ethiopia's Afar Triangle, where the African and Arabian plates of the earth's crust collided in a geological "hotspot." The ashes and boulder from 125 volcanic eruptions are used to date hominid fossils at sites in Ethiopia and northern Kenya.

Volcanic ashes are especially useful for providing a date for hominid sites, because these ashes go all over the place, while the rocks and lava of a volcano tend to stay near the source, arrested by gravity. Ashes that ride the wind are called aeolian. An ash found near the island of Lamu, off the eastern coast of Kenya, was matched to one at Koobi Fora. The same ash has also been found in Ethiopia's Awash Valley. A single volcano sent its dark billowing cloud over 385,000 square miles.

Ashes that accumulate in the landscape are called tuffs. In their most ideal form they ripple across an exposure like a gray ribbon. You can actually see the gray ribbon of tuff at the Zinj site.

Tuffs may be a few inches thin or 50 feet thick. They may settle down and be contained into neat layers by sediments, or they may be moved and redeposited by rivers. Their long line may be broken by faults, or buried at some points and exposed at others. Some tuffs are relatively pure, with shards of glass dazzling like diamonds. Others look like mud, sandstone or the bottom of a fire-

American geologist Frank Brown looks at a volcanic tuff, the light-colored line running along this exposure. (© Delta Willis)

place. Once a tuff is identified by a chemical "fingerprint," wherever that tuff is found, a geologist has an immediate reference for age.

This method of "fingerprinting" the tuffs was refined by American geologist Frank Brown, who refers to them as "thin slices of time." The fingerprinting method was just beginning to be developed during the KBS controversy. The tuffs at the KBS site were dated initially by the potassium/argon method. But part of the confusion was that geologists first thought there were only six tuffs at East Turkana. As it turns out, there were 70. This represented 70 different volcanic eruptions and 70 different dates.

Another problem is that the KBS site itself is what geologists call a "mess." There had been so many changes on this landscape that it is difficult to follow the levels of the earth from one point to another. When Harvard paleontologist Stephen Jay Gould visited the KBS site in 1986, he took a long walk around, trying to figure out the stratigraphy. (Professor Gould also teaches geology at Harvard.) He returned frustrated and said to Richard Leakey, "You're right; it's a mess. I mean there are no marked events that you can trace. I can't follow it."

"With all the things we know now, it's easy to say afterwards what went wrong," Richard Leakey allows in *The Hominid Gang*. "But people got emotional, attached and defensive." He includes himself among them. "It was a terrible time," recalls Frank Brown, "a terribly emotional business."

Geologist Brown then worked on the Omo team, as did Donald Johanson. It was the Omo team that initially raised doubts about the date. Ironically, in 1984, the age of Johanson's own discovery of Lucy was revised from four million to three million years old—a difference of one million years—the exact revision made on 1470. But no one said a peep about it. Why was the age of 1470 so controversial?

Discovered in 1972, 1470 remains the most complete and oldest skull of *Homo habilis*. It was an exceptional discovery; it made Richard Leakey famous, and its age supported the "Leakey line," that *Homo* went far back in time. Johanson's 1974 discovery of Lucy, even with the younger date of three million, was still an

Harvard paleontologist Stephen Jay Gould discusses the KBS tuff with Richard Leakey. Gould, who teaches geology, visited the fossil sites of Lake Turkana in 1986. (© Delta Willis)

older australopithecine. So what was the problem? Emotions and egos—none of which belong in a good scientific debate.

THE PIG CLOCK

The KBS controversy began in 1971 when paleontologist Basil Cooke did a study of pig teeth from three different areas: the Omo River Valley, Olduvai Gorge and Koobi Fora. There is a certain species of pig that evolved rapidly, and the changes it underwent could be seen in its molar teeth. Cooke had done such a good study that when a certain shape of that pig's tooth was found, the age of the site could be determined. This species of pig was so abundant that it could be used to double- check the geologists' work; today Cooke's work is referred to as the "pig clock."

Dr. Cooke found that the pig teeth from below the KBS tuff matched those from the Omo and Olduvai that were only two million years old. Thus, skull 1470 also should have been dated at two million years old. But, using the potassium/argon method, the lab at Cambridge had dated the tuff at 2.6 million years old. (Because the skull was found 40 yards below the tuff, it was older than the tuff, so the Cambridge lab estimated it was three million years old.)

To deal with the problem, the Koobi Fora team reckoned that the pigs were evolving at a different rate in different places. This reasoning turned out to be false, but for the moment, the team stood behind the results of the Cambridge lab. Over the next few years, scientists begin to choose sides—those that questioned the KBS date and those that didn't. This pitted the Omo team against the Koobi Fora team.

The composition of the sides of the debate tells us that it wasn't a real debate. If people were really trying to find out the truth, they wouldn't have divided up so neatly into two opposing teams. There would have been people on the Koobi Fora team who had some doubts and people on the Omo team who argued that the KBS date might be good. But the debate became like a football game—Us vs. Them—and the issue was who would win. The Omo team was out to prove Leakey wrong, and Leakey's team was on the defensive.

At a 1973 conference in Nairobi, the Omo and Koobi Fora teams met to review the evidence. Archaeologist Glynn Isaac, who worked with Richard Leakey's team, joked that the Koobi Fora team needed "pig-proof" helmets. The Omo team didn't laugh; they thought he was making fun of their work, when he merely meant to lighten things up and relieve some of the tension. This was the second clear sign that the debate was not going to be real or fruitful, because the scientists had lost their sense of humor—which is critical to accepting the notion that you might be wrong.

Isaac, his sense of humor intact, began to think that the KBS date might just be wrong. He wanted a second opinion. He talked to geologist Garniss Curtiss at Berkeley, who had done the accurate dating for the Zinj skull at Olduvai. After analyzing samples from the KBS site in his lab, Curtis discovered there were actually two tuffs at this level of the site. The oldest he dated at 1.8 million years old. His results suggested that the pig clock was right.

Curtis announced the results at a London meeting in 1975, where Basil Cooke gave an even stronger report on his pig teeth. Richard Leakey felt betrayed. He criticized Cooke for not showing him his paper beforehand, and he criticized Curtis for not talking to him directly about his work. He felt that Glynn Isaac had "totally abandoned the ship" by encouraging a second opinion on the date. And Mary Leakey remembers being treated with a "distinct coolness" by her son when she expressed her own doubts.

All of this time, Richard Leakey was being assured by the Cambridge lab that the older date was good, that they had done several more tests confirming this. As Richard expected people on his team to be loyal, he was loyal to the Cambridge lab and felt he should defend their work. While his loyalty was understandable, it was a mistake, because the Cambridge lab turned out to be wrong.

Richard also made a mistake by not encouraging the research of Curtis and Cooke. Paleontologist Tim White accused Leakey of making another error by "censoring" White's paper on pigs' teeth. Leakey denies this, but White left Nairobi very upset and angry, and teamed up with Richard's leading critic, Donald Johanson.

With all of those mistakes, some lessons were learned. In 1980 Richard Leakey invited geologist Frank Brown to work with the Koobi Fora team. Brown was totally surprised by the invitation. "I was in the Omo bunch, and we were on the other side of the KBS controversy. To have this invitation to come finish the geology at Koobi Fora just floored me." Several years later Brown asked Richard, "Why me?"

Richard told him, "For one thing, your arguments turned out to be right, and that's good. Also, your arguments were never personal. It's fair enough to argue about science, but you never joined the character assassination."

Frank Brown's dates for the hominid sites of Kenya are the most accurate ones in the field today. Because the volcanic ashes that Brown identified stretched into Ethiopia and Tanzania, sites there have been dated accurately as well. Some of the ashes that settled in Tanzania preserved one of the most extraordinary discoveries ever made about our ancestors.

LAETOLI

Twenty miles east of Laetoli, a volcano known as Sadiman erupted 3.65 million years ago. As the ashes settled to the ground, there was a series of rain showers.

Following one of these showers, a group of hominids walked in the Laetoli area. The ground was still wet and mushy, so they left their footprints behind them. When the sun came out and baked the ashes, these footprints were preserved, as hard as a fossil. Over the years, the winds brought in new sediments, and layers of earth covered many of these footprints. Some, however, were exposed by rains again recently. The first footprints to be found belonged to hares and to birds similar to guinea fowl.

While the KBS controversy was brewing in the mid-1970s, Mary Leakey devoted herself to exploring the Laetoli region. Much of her work at Olduvai had been completed, and she moved to this older site nearby in hopes of finding older stone tools. At the time, the oldest known stone tools were from the Omo, dated at 2.2 million years old. But Mary and her team had found older hominid fossils at the Laetoli sites, and the big question was, did these hominids make tools? As Mary has said, "In archaeology you almost never find what you set out to find."

The way the first footprints were found is as remarkable as it is funny. In September 1976, Mary welcomed a group of friends who had come down to Tanzania for a quick visit. The group included Kay Behrensmeyer, Andrew Hill, a paleontologist who worked for the Nairobi Museum, and Jonah Western, an ecologist who worked at Kenya's Amboseli National Park. Mary was showing them the sites one day, and as they walked back to camp in the evening, Hill and Western began to toss elephant dung at each other. Elephant dung, when dried, is not offensive at all; it doesn't smell; it's just like a big cake of dried grass. Western hurled a big piece at Hill, who ducked and fell on the ground.

Hill noticed some interesting-looking imprints in a flat gray surface. The first little dents he saw were later identified as raindrop prints. But it made him look around, and he found footprints of hares, birds and rhinos. Since then literally tens of thousands of footprints have been found in what became known as Site A, ranging from the tiny tracks left by insects to the massive depressions left by elephants.

Two years later a footprint was found that looked like the mark of a human heel. Excavations began, and it turned out that the path of footprints went for over 20 yards. It looked as if two different hominids had walked there. Mary Leakey thinks they didn't walk at the same time. The path of one individual was very clear and distinct, while the other set of footprints was blurred and not as sharp. It appeared that they may have been impressed upon two different layers of ash which fell at different times.

Mary Leakey describes the Laetoli footprints as "perhaps the most remarkable finds I have made in my whole career." Because the footprints were so humanlike, she felt they could only have been left by one of our direct ancestors.

What do these footprints tell us about our ancestors? That they walked upright 3.65 million years ago. The shape of the foot is very similar to our own. Because no tools were found in this same level, these hominids must have walked upright before they used tools.

In addition to the footprints, Mary's team found part of a child's skeleton and several fossil remains of adults—two lower jaws, part of an upper jaw and a number of teeth. The best specimen of the lower jaw was called LH 4, for Laetoli Hominid number four. It would become a huge bone of contention.

BONES OF CONTENTION

After studying these hominid fossils, Dr. Leakey felt that these Laetoli finds belonged to an ancient form of *Homo*. She based her judgment not only on the details of the bones; she also felt that the humanlike footprints confirmed this. Paleontologist Tim White, who worked on her team at Laetoli, prepared the scientific report on the fossils. He too described them as a species of *Homo* in a paper that was published in the British journal *Nature*. Everyone agreed that the fossils from Laetoli looked very similar to those found by Don Johanson's team in Ethiopia a year after Lucy was discovered. The second Ethiopian collection was known as the First Family, because it included remains of several individuals.

They were very different in size and details from the australopithecine Lucy.

Johanson also thought that the First Family from Ethiopia was *Homo,* just as Tim White thought the Laetoli hominids were *Homo.* This belief supported the Leakeys' theories. But after White had his disagreement with Richard Leakey and teamed up with Don Johanson, he changed his mind about the genus the fossil belonged to, and then he changed Johanson's mind.

White and Johanson lumped all the fossils from Ethiopia and Laetoli together and decided they were *all* australopithecines. Mary Leakey didn't agree with this. Nor did Richard, who always thought that there were two species among the Ethiopian finds—an australopithecine and an older form of *Homo.*

In 1978 Johanson and White decided to announce a new name for the discoveries. Johanson was one of many scientists scheduled to speak at a Nobel Symposium in Sweden in May. The conference would honor Mary Leakey, who would receive a medal from the King of Sweden for her scientific work.

Mary Leakey received the Golden Linnaean Medal, the first woman to do so. But she also endured one of the most embarrassing moments in her life. Johanson spoke before she did, and announced the new name for the species from Ethiopia—and in this species, he included Mary Leakey's discoveries from Laetoli. In fact, the jaw LH 4 was featured as the type specimen, or model, for the news species name *Australopithecus afarensis. Australopithecus* is the formal genus name for the australopithecines, and *afarensis* denotes the Afar Triangle of Ethiopia where they were found. But the model specimen came from Tanzania.

Johanson talked at length about the discoveries at Laetoli, which scooped Mary's own speech. She was angry and embarrassed. During the coffee break, she confided to Richard: "How am I going to give my paper now? It's all been said." Richard said later that she felt that she "was going to look as if she was a fool, repeating the material." More controversial, Johanson had named her discoveries, using a designation that was totally at odds with what she believed. Because Johanson named them first, that name stuck. When she stood up to give her talk, Mary

Leakey could not say that the finds from Laetoli were *Homo* as she thought they were. She just expressed her deep regret that "the Laetoli fellow is now doomed to be called *Australopithecus afarensis.*"

Some scientists suggest that White and Johanson lumped the Laetoli finds in with the others to give the new species an older date. The fossils that Mary Leakey found in Tanzania were nearly four million years old—at the time, the oldest hominids ever discovered.

The controversy continues today. In 1990 more fossils were found in Ethiopia that suggest there may have been two species, rather than one as Johanson and White claimed. But the Leakeys withdrew from this debate; they decided that their best defense was to just keep working and find more evidence.

CHAPTER 8 NOTES

p. 89 "It's no good being in front . . ." Raymond Dart as quoted in John Reader, *Missing Links*, p. 94.

p. 91 "the organization I had built up . . ." Richard Leakey, *One Life*, p. 196.

p. 92 "first real and successful challenge to the Leakey dynasty . . ." Donald Johanson and Maitland Edey, *Lucy*, dustjacket.

p. 94 "thin slices of time . . ." Frank Brown as quoted in Delta Willis, *The Hominid Gang*, p. 175.

p. 94 "You're right; it's a mess . . ." Stephen Jay Gould as quoted in Delta Willis, *The Hominid Gang*, p. 163.

p. 94 "It was a terrible time . . ." Frank Brown as quoted in Delta Willis, *The Hominid Gang*, p. 188.

p. 97 Isaac "totally abandoned the ship . . ." Richard Leakey as quoted in Roger Lewin, *Bones of Contention*, p. 234.

p. 97 "distinct coolness . . ." Mary Leakey, *Disclosing the Past*, p. 148.

p. 97 "I was in the Omo bunch . . ." Frank Brown as quoted in Delta Willis, *The Hominid Gang*, p. 188.

p. 98 "In archaeology . . ." Mary Leakey, *National Geographic*, November 1985, p. 599.

p. 99 "perhaps the most remarkable . . ." Mary Leakey, interviewed for *The Making of Mankind* television series, BBC Television, September 4, 1979.

p. 100 "How am I going to give my paper now? . . ." Mary Leakey at the Nobel Symposium and Richard Leakey response, in Roger Lewin, *Bones of Contention*, p. 269.

9

THE TURKANA BOY

In 1980, after a dozen years of research at Koobi Fora, Richard Leakey decided to move his team to the west side of Lake Turkana. He had found some older exposures there, including rich Miocene sites, and Koobi Fora had already provided many extraordinary discoveries.

The fossil remains of more than 200 early humans had been unearthed at East Turkana, including eight nearly complete skulls. This evidence proved that three different kinds of species had lived in the same area at about the same time. They included the large-brained Handy Man like 1470, the small-brained Zinj and a third species that Leakey has not given a name. The third species is represented by two skulls known simply as 1813 and 1805. They are mystery skulls that remain in what's called the "suspense account." After all of the controversy, Leakey wanted to be cautious about assigning a new name.

New discoveries in paleoanthropology often undo older ones. These new discoveries may be very small—such as adding the "turtle" bones to the back of the *Proconsul* skull. But West Turkana was to produce several big, profound finds, including the most complete ancient skeleton of our ancestors ever unearthed.

There the Leakeys, Alan Walker and the Hominid Gang found a whole skull and plenty of teeth. They found arm bones and leg bones and even delicate shoulder blades that, when held to the sun, admit light. The teeth revealed that it was a teenager, the pelvis suggested that it was a male, the leg bones confirmed that it walked upright—the stance for which *erectus* is named.

The full forehead and a round, smooth braincase indicate that this skull held a brain much larger than those of the australopithecines. As mentioned, a large brain is the defining characteristic of *Homo*. In a search where controversy about names has brewed for decades, no one challenged the name for this find. It was a *Homo erectus*. It was the oldest *Homo erectus* ever found. It is called the Turkana Boy, in honor both of the name of the vast lake and the nomadic people who still live along the western shores.

A SKELETON EMERGES

The first clue to the skeleton was discovered in 1984 by Kamoya Kimeu, the leader of the Hominid Gang. Richard Leakey was back in Nairobi, working at the museum as he did on most weekdays. His job as museum director included supervising several departments—in archaeology, paleontology, botany, primate studies and educational programs. He had just established a new center for prehistory, a memorial building named in honor of his father. He is up before dawn and in his office by 6 A.M. The search for hominids is only a small part of his duties, and he delegates much of the search to teams that work in the field throughout the dry season, from August to March. He and Meave often fly up to the sites on weekends to work and to see what has been found. Meanwhile, the search goes on without them.

During the early part of August in 1984, Kamoya Kimeu had been leading a team of six of his gang to search for fossils, working alongside paleontologist John Harris and geologist Frank Brown. For several weeks they had been looking for bones on the west side of the lake, north of a river called Nariokotome (na-rée-uh-cót-uh-mee).

They hadn't had much luck finding any hominid bones in the north, and on the evening of August 12, the three men met on the banks of the river to discuss what to do. They thought there must be hominid bones farther south; perhaps they should break camp and move that way soon.

At dawn the next day, Kamoya hooked up the radio telephone to the battery of the Land Rover and called Richard at the museum in Nairobi. Richard approved the move south, but because the men had been working for 14 days without a break, he suggested they take a day off to rest. Most of the men wrote letters, did their laundry and slept. Kamoya decided to go for a walk.

He walked directly south, across the Nariokotome River, which was bone dry at the time. He took a stroll up a little hill dotted by Salvadora and acacia trees. Naturally he looked for fossils; if Kamoya is awake, he looks for fossils.

Richard once felt that he wasn't finding any fossils because he had spent the previous two weeks diving for lobsters in the Indian Ocean. "I realized my subconscious was making me look for lobsters. The next day I made it very clear to myself that the image in my mind had to be of bones and by this conscious effort I was soon seeing interesting fossils again." This is called a "search image." If you have it in your head to look for one thing, you might not even see something else.

Anyone can have a search image—if you are concentrating on finding blueberries, you may not notice the poison ivy, and vice versa. But scientists at work in the field are especially sensitive to this phenomenon. For example, when Stephen Jay Gould went fossil hunting with the Leakeys, he found lots of snails (or shells of snails). Meave Leakey was amazed. "How do you find them?" she asked. Gould said, "Oh, it's my search image." Professor Gould studied snails, and therefore they are his focus. This focus is just the brain's way of organizing all that we see into some order. A search image becomes hard with experience, creating a visual image in the brain's memory. Richard's memory had held onto the image of a lobster because of his recent experience.

Kamoya 's fossil search image made him focus on the ground, rather than looking at the birds in the trees or the camels on the horizon. He had wandered only 300 yards south of the camp when he spied a bit of fossil that had washed to the surface by erosion. It was a fragment less than an inch and a half long and not quite as wide. You could carry it in your pocket for days and forget it was there.

Not only was this fragment small, but it was the same color as the lava rocks around it. Kamoya held the fragment between his big fingers, feeling its curve and thickness. Then he rubbed his thumb on its smooth inner surface, feeling none of the texture found in other animal skulls. From experience, he knew immediately that this was part of a hominid skull.

The next day at dawn, Kamoya hooked up the radio telephone again and told Richard what had been found. Richard flew up in his plane with anatomist Alan Walker, who had been doing some work in the museum.

They started by removing some boulders from around the site. Then they began to sieve the topsoil—sifting it through a rectangular box with a screen, like a window screen—which produced nothing. The odds were against their finding anything else, and sieving is dirty, boring work. Richard points out that many important fossils are found when sieving is required; people will wander off and do anything—including discover something—to get out of sieving! Nonetheless, Richard picked out an acacia tree a few yards from where the fragment was found and suggested "If we don't find anything else by the time we reach this tree, let's quit."

For some reason, Richard chose to focus on the tree itself, digging around the roots. He plopped down a green cushion and used a big metal pick with a wooden handle, a paintbrush and and his fingers. After awhile he took off his shirt, hung it on the branches of the tree and lay down with his chest on the cushion, putting his face directly over the roots. Within a few minutes, he switched to a dental pick and began to make long, dry scratches against fossilized bones. He shouted for the others to come over and look. He had found the face bones of the Turkana Boy. The roots of the acacia had worked their way through the facial bones and halved the upper jaw. How did this happen?

Alan Walker suggests the tree grew out of the skull. The skull would have held moist soil, and in the area's semiarid environment, an acacia seed found it an ideal place to germinate. As the tree grew, the roots broke through the skull and slowly pushed it apart.

Now the men were anything but bored. They continued to dig, for days and weeks on end. They found rib bones in rocks and teeth in swampy sediments. How did they know the teeth belonged to this skull? They were from the same time level, and they were all the same age. And the teeth indicated that this boy was about 12 years old: He had no wisdom teeth, for example. Kamoya found the first tooth, then Alan Walker found a "P3," or third premolar. Once they cleaned the calcite off the teeth, they fit into the jaw; so there was no doubt that they belonged. Also, despite all of that digging and sieving, they didn't find any extra arm bone, no extra tooth. So the knew they were dealing with just one individual.

While Leakey and Walker worked on excavating bones with dental picks, the rest of the team made measured, slow cuts into the hill, looking for more bones. In an area measuring four yards by five yards, they found the back of the skull and the lower jaw, and much of the skeleton. There are 206 bones in a skeleton such as mine and yours; the count is the same on ancient hominids—the bones just differ in form.

There wasn't a great deal of difference between this boy's bones and ours, but because the find was so old and so rare, they wanted every bone—even the tiny fingertips. So they kept on digging and sieving for four long field seasons, removing 1,000 tons of soil. Every bit of soil was sieved and washed.

Richard and Meave Leakey visit a Miocene site at West Turkana, where members of the team sieve in search of bones. (© Delta Willis)

As the team worked with their big metal picks, they swept the earth to the side, where it was collected into large buckets. This earth was taken by Land Rover to the lakeshore, where it was dumped onto a sieving screen and carefully dipped into the water. The washing team, which included the Leakey daughters, pinched every little clod of dirt in hopes of finding a fingertip or a tooth. Louise Leakey, born in 1972, and Samira, born in 1974, joined the expeditions at a very early age. They continued working as part of the team during their school vacations.

Most of the bones were laying right on top of a layer of volcanic ashes known as the Okote tuff. Frank Brown determined that this tuff was about 1.65 million years old. Because the skeleton was just above it, the Turkana Boy was estimated to be 1.6 million years old. No Homo erectus that old had ever been found before. In fact the most complete skeletons of our ancestors found before this were 1.5 million years younger—the Neanderthals who lived in Europe.

What did the bones of the Turkana Boy tell the team? When Alan Walker reconstructed the skull it had one piece that he couldn't quite connect by virtue of grain or shape. "It had a big hole, a pit so big you could put your thumb into it." It turned out to be the bony socket known as Broca's area, the part of the brain that manages speech. And this boy was very tall, what Leakey referred to as "strapping." He was five feet four inches when he was 12 years old. Lucy and some of the other australopithecines were very short by comparison.

The discovery made the major news magazines, and it also made Kamoya Kimeu famous. Richard Leakey wanted to make sure that credit was given where it was due. Kamoya had been finding bones for 30 years.

In the 1950s Louis Leakey had invited Kamoya to Olduvai Gorge to work. The Leakeys provided food, shelter, salary and instructions. "First they taught us to know fossilized bones," Kimeu said. "You know, you might think it is only a stone. Then they showed us animal bones, then primate bones and very slowly, hominid bones." Kamoya Kimeu was the one to find the Peninj jaw at Lake Natron, and he found so many fossils at Olduvai that the Leakeys named one of the finds, a new species of monkey, in his honor: *Cercopithecus kimeui.*

Kamoya Kimeu, the leader of the Hominid Gang, who found the first clue to the Turkana Boy. In 1986 Kimeu was honored for his discoveries by the National Geographic Society.
(© Delta Willis)

To honor his work, Kamoya was awarded a special medal by the National Geographic Society. President Ronald Reagan made the presentation in the Oval Office of the White House in 1986. Afterward, with Richard's help, Kamoya endured his first press conference, as described in *The Hominid Gang*. He explained that he doesn't study bones; he only finds them. "Mostly it's a jaw here, a piece of skull there." Of the Turkana Boy, he said, "That was the big one."

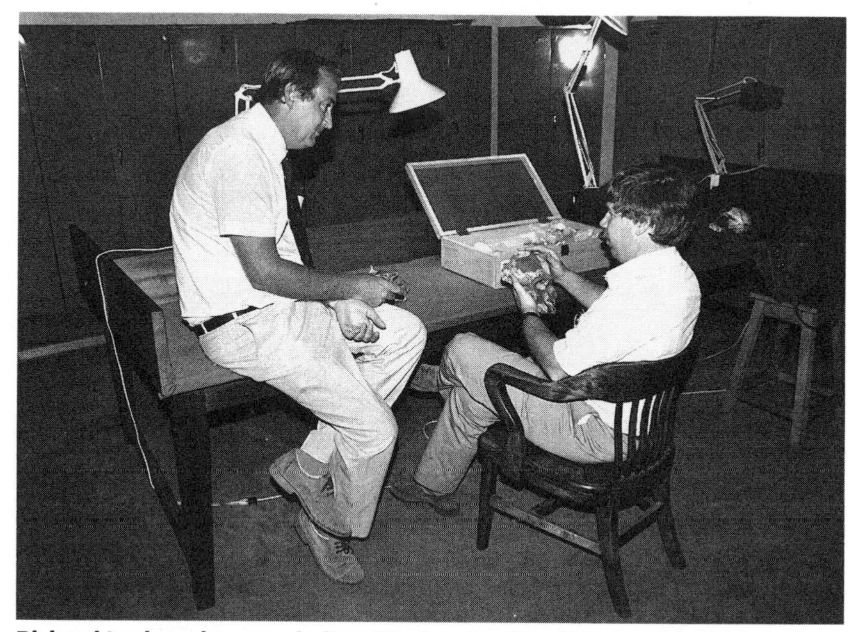

Richard Leakey shows a skull to Stephen Jay Gould in the Hominid Vault in Nairobi. All original fossils are kept in this air-conditioned room at the Louis Leakey Center of Prehistory, where scientists from around the world come to study them. (© Delta Willis)

Today, a cast of the Turkana Boy can be seen in the Discovery Hall at the National Geographic Society building in Washington, D.C. The original bones are kept in the archives of the Nairobi Museum, in a special bombproof, air-conditioned room known as the Hominid Vault.

The Hominid Vault is also home to the next big discovery from West Turkana, found in 1985 by Alan Walker. It was such a surprise that it would cause all the scientists to agree that it was time to go back to the drawing board: They had to reconsider the branches on our family tree.

OUR FAMILY TREE

As you can tell by the discoveries described in this book, our knowledge of the human family tree is always changing. Often, throughout the history of the science, ideas about a single discovery are changed when a new discovery is found. This happened with Zinj, when the Handy Man was found only a few years later. It happened with the Piltdown Hoax. In fact, if you look at textbooks published only a few years ago, you might find drawings of a primate called Ramapithecus, described as being the oldest hominid. Only in 1982 was it discovered that Ramapithecus is an ancestor to the orangutan, and not us. The scientists who had made their claims about Ramapithecus were not as careful as they should have been in their studies, yet they pushed their ideas, going so far as to say that this primate probably hunted and walked upright.

Some paleoanthropologists make claims that are simply not yet proven. Why do they do this? Some feel a great deal of pressure to make the claims to that they can obtain more funding to support continued research. Some also seek the publicity and fame. The most effective way to receive publicity is to suggest that your discovery proves someone else's wrong; it helps if that someone is famous. It's not good science, but some of the press and the public love it.

In 1981 Richard Leakey was invited to appear on a CBS program in a series called Universe, hosted by Walter Cronkite. Leakey was told that the discussion "would center on the process of anthropology and how we know what we know about our earliest ancestors." Understanding evolution was a problem then, as it is now. A survey of college students found that many believed cave people battled dinosaurs—creatures that disappeared 60 million years before the first early hominids were on earth. Also, religious fundamentalists were trying to censor any mention of evolution in school textbooks and to have their religious views taught under the guise of "creation science." Several court cases, including ones in Arkansas and Louisiana, were similar to the famous Scopes "Monkey Trial" held in Dayton, Tennessee in 1925, in which a schoolteacher was tried for teaching Darwin's theory in class.

The Universe program was taped at the American Museum of Natural History in New York. When Leakey arrived, he was surprised to see Donald Johanson on the stage and the rest of the scientists from the museum sitting in the audience. Richard had thought there was to be a panel discussion among several leading scientists. He preferred not to debate with Johanson, who had written a book that criticized the Leakeys' work and promoted his own theories about a small detail of the human family tree. Richard had co-authored a paper in response that had appeared in a scientific journal, and in this instance he recalled, "I don't think it helps the public understand evolution to focus on a disagreement. The search has nothing to do with Johanson verses Leakey; there are many people involved. . . . It's according to how you look at the fossils. But [the question of a common ancestor] is not an issue fundamental to understanding evolution."

Richard felt it was wrong to emphasize controversy. But as soon as Mr. Cronkite began to introduce the two men according to his script, Richard knew he was in for a debate about Lucy. "Before the discovery of Lucy made Donald Johanson a celebrity," Cronkite said, "the king of the mountain of paleoanthropology was Richard Leakey." A program that was supposed to be about science, the how they know what they know, turned into one about disagreements over details. As Stephen Jay Gould pointed out (in a 1986 interview by the author), "You might have thought there was a dramatic change in the family tree—like we were related to hyenas instead of apes!"

As the cameras continued to tape, Leakey became angry: "I realized that I'd been absolutely set up." Johanson talked about his view of the family tree. To emphasize their disagreement, Johanson pulled out a chart with a drawing of his version of the tree on one side and a blank space on the other side for Leakey to draw his version. Richard hesitated at first; then he took a pen and put a big X across Johanson's tree. "And what would you draw in its place?" Johanson asked, slightly taken aback. Richard then drew in a huge question mark in the blank space.

Johanson had described his chart as "the family tree that Dr. White and I have established." Four years later, the discovery of the black skull changed that tree.

THE BLACK SKULL

In August of 1985, when Alan Walker found the first clue to the black skull, he wasn't looking for a hominid. He was looking for a monkey.

The area known as Lomekwi is an hour's drive south of Nariokotome. The skull of a hippo had been found there a year earlier, as had the arm bone of a monkey. The Hominid Gang had left the bones in place, or *in situ*, until their position could be marked on aerial photographs. As he walked back to the hippo site at West Turkana, Alan noticed another small fossil on the way. It was only a small, dark fragment. He picked it up—gave it a look—then put it back down. Then he picked it up again.

As he studied it more and began to look around, he was sure that he had found parts of a hominid skull. He eventually strolled over the hill to the hippo site, where his wife, Pat Shipman, was working. Walker, British-born, has a deadpan sense of humor. Having just stumbled across what would be described as "the greatest fossil find since Lucy," he very casually said he had a hominid bone to show her. "Great!" Pat exclaimed. "What part?" He said, "Oh, a skull." It had taken 10 years of searching to find eight good hominid skulls at East Turkana.

In 1986, after Alan and Richard had gathered all the pieces, put them together and made their report, the *New York Times* ran a story with a headline "New Skull Finding Challenges Views." The *Times* report went on to say that the discovery "may shake some old branches on the family tree."

The discovery is called the black skull because as it fossilized, it took on the color of the surrounding dark minerals, which gave it a manganese hue. It looks like a mean version of Zinj, with enormous teeth, bold brow ridges and a massive crest on its small braincase.

"People didn't expect to find an ancestor of Zinj looking like that," Alan Walker said shortly after its discovery. When the discovery was announced, paleontologists were surprised by its age—it was 2.5 million years old, nearly a million years older than Zinj. "Whichever way you look at it," Dr. Fred Grine noted in *Science* magazine, "it's back to the drawing board."

"Kamoya's boy was very easy for the public to understand," Walker said. "All you had to say is: See, here's a complete skeleton. But the black skull—it's nitty gritty taxonomy and science." The nitty gritty was whether the black skull was an older "Zinj," or whether it was a new species. But it proved one thing: a branch on our family tree that had been drawn in a single line now needed

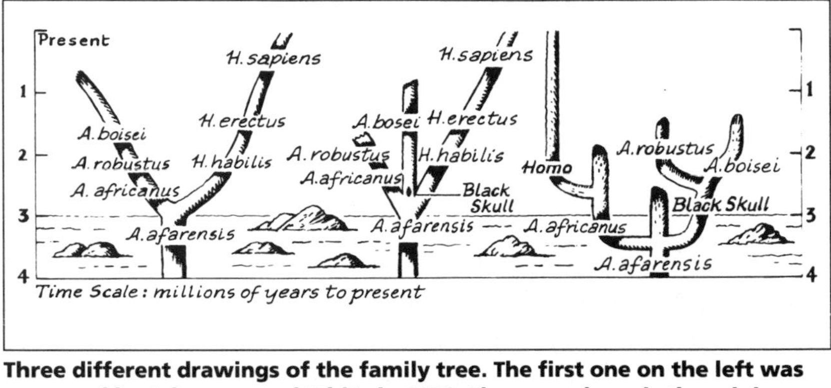

Three different drawings of the family tree. The first one on the left was proposed by Johanson and White in 1979. The second one is the minimum revision forced by the 1985 discovery of the black skull. The third one on the right is one alternative proposed by Eric Delson, a paleontologist at the American Museum of Natural History. (Bob Gale)

to be split into two or more branches. (See different versions of the tree above.)

Because so little evidence exists, scientists often compare finds from distant sites. Finds from China, for example, helped explain those in Europe and Africa. Richard Leakey suggested that the black skull proved there were actually two species at Hadar in Ethiopia, and the skull was one of these. This idea came from the remains from the very back of the skull—the same area where the two "turtle" bits fit into *Proconsul.* The black skull has a bony crest that extends back there, but the face looks very different from the reconstructed skull of *A. afarensis.* (None of the other adult finds from Hadar had a whole skull, so one was put together using bits from 12 different individuals.) Leakey thinks the reconstructed skull may have included bits from two different species. Johanson's team doesn't think so, but more evidence is needed to resolve the issue.

In 1982 the government of Ethiopia banned Donald Johanson and his team as well as all other foreign paleoanthropologists from working in Ethiopia. The ban was lifted five years later, but meanwhile, in 1986, Mary Leakey having retired from fieldwork, Don Johanson and his team began to research at Olduvai Gorge. The results of their research are described in a book, *Lucy's Child.*

THE END OF AN ERA

The Cronkite program had made the search for human origins seem like a contest to find the "oldest" ancestor. If that was the issue, none of the discoveries of the two men featured on the program would have qualified. The oldest known ancestors come from Laetoli, Tanzania, where they were discovered by Mary Leakey and her team. Actually, Mary Leakey found the earliest human twice—Zinj was the oldest known hominid when she discovered it in 1959, and A. afarensis is the oldest known hominid today.

She did not name "the fellow from Laetoli" *afarensis,* nor did she name Zinj. She is not known for her version of the family tree, but for the many discoveries that helped fill in the branches, namely: *Proconsul,* Zinj, Handy Man and *afarensis.* And Mary Leakey and her team are known for discovering the oldest evidence of our upright gait, the footprints of Laetoli.

Mary Leakey's long tenure in the field has allowed her to see the family tree change many times. The branches for Zinj and the Handy Man didn't exist when she began. Mary Leakey worked at Olduvai Gorge over nearly 50 years—from 1935 to 1984.

In August of 1982 she suffered a hemorrhage that left her blind in one eye. At the time, she was nearly 70 years old. Her family was concerned about her health, especially because she was so far from any hospital or medical help at Olduvai. In 1984 Mary gave up her fieldwork in Tanzania, to move back to the Langata house in Kenya. "My field days are basically over," she said. "I may do a little excavation, but now I must concentrate on all this writing." She spent countless hours preparing the detailed reports on all of her finds from Olduvai and Laetoli—the thousands of stone tools, the thousands of bones. She modestly referred to this as "the consuming part of my work."

She also helped piece together the many new *Proconsul* finds from Rusinga, and lectures on tours in Europe and the United States, especially in association with the National Geographic Society and the L. S. B. Leakey Foundation. In 1983, she published *Africa's Vanishing Art,* on the Stone Age drawings of Tanzania. In 1984 she published an autobiography entitled *Disclosing the Past.*

As of 1991, she continues to work, both at the Nairobi Museum and out of her home, where she is still surrounded by Dalmatians.

Stephen Jay Gould has referred to Mary Leakey as the "unsung hero" of the search for human origins. Andrew Hill said, "Mary's importance is often underestimated, particularly by younger members of the field, because many of the techniques and methods she introduced are now commonplace."

Her legacy is such that a London shopkeeper glanced twice at a check she had written to purchase a gift, and then blurted, "Oh! Are you the Prehistoric Mrs. Leakey?" Mary laughed and said, "I suppose so."

CHAPTER 9 NOTES

p. 105 "I realized my subconscious . . ." Richard Leakey, *One Life*, pp. 139–140.

p. 106 "If we don't find anything . . ." Richard Leakey, Alan Walker, and Kamoya Kimeu on the discovery of the Turkana Boy as quoted in Delta Willis, *The Hominid Gang*, pp. 236–239.

p. 108 "First they taught us . . ." Kamoya Kimeu, interview with the author, 1986.

p. 111 "would center on the process of anthropology . . ." Letter from CBS "Universe" producer as quoted in Delta Willis, *The Hominid Gang*, p. 303.

p. 112 "I don't think it helps . . ." Richard Leakey as quoted in Delta Willis, *The Hominid Gang*, pp. 303–304.

p. 112 "Before the discovery of Lucy . . ." Walter Cronkite, quoted with permission of CBS News, in Delta Willis, *The Hominid Gang*, p. 306.

p. 113 "the greatest fossil find since Lucy . . ." John Noble Wilford, "New Skull Finding Challenges Views," *New York Times*, August 7, 1986.

p. 113 "People didn't expect . . ." Alan Walker, interview with the author, 1986.

p. 113 "Whichever way you look at it . . ." Fred Grine, "New Fossil Upsets Human Family," *Science*, August 15, 1986, p. (n/a).

p. 113 "Kamoya's boy was easy for people . . ." Alan Walker, interview with the author, 1986.

p. 115 "My field days . . ." Mary Leakey, interview with the author, 1986.

p. 116 "unsung hero." Stephen Jay Gould, *Ever Since Darwin*, p. 56.

p. 116 "Mary's importance is often underestimated . . ." Andrew Hill on Mary Leakey as quoted in Delta Willis, *The Hominid Gang*, p. 92.

p. 116 "Oh! Are you the Prehistoric Mrs. Leakey?" quoted in Delta Willis, *The Hominid Gang*, p. 90.

10

THE LEAKEY LEGACY

"As a family the Leakeys have always relished challenges," Richard Leakey wrote in his autobiography. This partly explains why they became legends in their own lifetimes.

In October of 1984, the *New York Times* ran a two-page article with the headline: "The Leakeys, A Towering Reputation." The *Times* story mentioned that "Even bitter rivals find room to praise the family's accomplishments." In a field full of controversy and envy, that is a great compliment. As mentioned in Chapter 1, *Science Digest* referred to the Leakeys as the "first family" of paleoanthropology. While the Leakeys often worked on separate research projects in different locations, it is unusual for one family to be such a dynamic force within a science.

Each has been a leader at the top of his or her field. Mary Leakey is considered the leading archaeologist in the world of prehistory. But she is also an excellent artist, and she has made some of the world's most spectacular fossil discoveries. Beyond stones and bones and art, she and her team found the incredible footprints of Laetoli. Any other scientist in this field would envy just one of her discoveries, but it seems each of the Leakeys have had several careers, and excelled in every one of them.

Louis Leakey was intrigued by stone tools at a very early age, and he helped to find many of the famous fossil sites in East Africa. He was also a dynamic public speaker and inspired many students, including the world's most respected primatologist, Jane Goodall. Louis also helped to develop the Nairobi Museum and established a primate center. Richard has found many important fossils and was described as the "organizing genius

Richard and Meave Leakey work together at a Miocene site at West Turkana, where three unknown species of apes were discovered in 1986. Meave is especially skilled at piecing fossil fragments together. (© Delta Willis)

of modern paleoanthropology." He has written several books and hosted his own television series, *The Making of Mankind*. Now he is even more widely known as one of the world's leading conservationists.

In addition to liking challenges, another part of their success has to do with the idea of an extended family. They seem to inspire each other, and everyone who works alongside them—such as Kamoya Kimeu. Richard's two brothers, Jonathan and Philip, also made fossil discoveries when they were young. As mentioned, Jonathan became an expert on snakes and runs a successful safari lodge. Philip is the only white member of the Kenya parliament and serves as a government minister. Richard's wife, Meave, is an accomplished scientist who is a leader in her field of paleontology, fossil monkeys. Meave's brother-in-law, John Harris, is the leading expert on fossil giraffes.

The Leakeys have been inspired, hard workers who often dared to venture where other scientists didn't. Louis Leakey was brave enough to look for the bones of our ancestors in Africa when many other scientists were looking in Europe and Asia.

They have made mistakes, but mistakes are an essential part of science, for the whole process involves trying to prove something wrong to find out what is right. Mistakes are also inevitable when you are bold and passionate.

Louis *was* bold and passionate, and often rushed to conclusions. Mary recalled that he would get ideas about where fossils might be and "would go off for no good reason. He was very often right, but very often wrong as well." When asked about her first impression of Louis, she said, "I thought he was a very vital sort of person, full of energy."

On the other hand, Mary is quiet and reserved. As a scientist, she is always careful and cautious. "If we had both been the same kind of person," she said, "we wouldn't have accomplished as much." The same sort of partnership is true of Richard and Meave Leakey.

Meave Leakey is the head of paleontology at the Nairobi Museum. While she helps a great deal with the hominids, her area of study is fossil monkeys. She is widely respected as a careful and thoughtful scientist.

Meave's ability to piece together fragments is legendary; she helped assemble the more than 300 pieces to skull 1470, and when she works on finding the right bone to go in the right place, Richard defers to her, saying "I'll leave this to you." Before presenting his ideas in a paper or in a lecture, Richard reviews them with his wife first.

"She's had very good scientific training" at the University of Wales, Richard notes, "and she's much more down to earth as opposed to my flights of fancy. She acts like a screen, a sieve, and has a tremendous influence on my scientific work." Richard describes Meave as "the most central and important part of my life and operation in many ways." While she often compromises her own studies to help Richard with the hominids, the fossils sites he finds in Kenya provide her with great and varied research material.

Meave first ventured to Africa in 1965, in response to an ad in a British newspaper that Louis Leakey placed for a research assistant at the primate center. In the past she has been very shy of the spotlight, explaining "It's not the credit I want; I enjoy the work." But Like Mary, she took to public speaking eventually, and recently began to lecture in the United States, sharing the podium with Alan Walker. She and Walker make a good team; in the field, they sometimes race to see who can find the bones that fit into place during a reconstruction. As it happened, they played the same game as children. When bored with a jigsaw puzzle, they both turned the pieces upside-down to the blank side, gaining good practice for piecing together fossils. Now they use these skills in friendly competition.

Richard was openly competitive with his father, whom he describes as "a strong man . . . a powerful man." While the two disagreed on professional matters over the years, and at times "We were a bit tense," in hindsight Richard views his father as a mentor.

"I greatly enjoyed his ability to inspire people. I try to pattern what I do in public on the same thing—being willing to talk to children and professors, and simply give of myself to others—I saw so many people go so far on his words. I think his very wide interest in so many things was also of great importance to my character formation."

Both Louis and Richard were active conservationists long before it became fashionable. Louis helped to establish some of the national parks in Kenya and wrote a book on the animals of East Africa for the National Geographic Society. Richard served as chairman of the East African Wildlife Society and helped found the Wildlife Clubs of Kenya, teaching young people the value of conservation; the successful program has since been adopted in several other African countries.

African children have quite a different view of wildlife from students in the Western world; few of them can afford to travel to the national parks, and their encounters with wildlife around their homes are often negative ones, because baboons and elephants can destroy their parents' crops. The purpose of the Wildlife Clubs is to ensure that the future leaders of Kenya find value for their natural heritage. The exhibits and displays at annex museums around the country also ensure that young Kenyans will appreciate another aspect of their heritage—this landscape as the setting where all humans began.

Over the years, several talented Kenyan and Tanzanian students have pursued careers in paleontology and archaeology as a result of the Leakeys' support. Scholarships to study at universities in

Europe and the United States were provided by Richard's organization, the Foundation for Research into the Origins of Man (FROM), and the L. S. B. Leakey Foundation. In 1984 the base camp at Koobi Fora was turned into the Koobi Fora International School of Archaeology and Paleontology, which operates in association with Harvard University. Every summer college students go to East Turkana to learn how to find and identify fossils and tools, while gaining college credits. In the summer of 1990, one student found part of a Zinj skull.

Richard also insists that every foreign researcher who goes to Kenya must employ a qualified Kenyan student as part of the team in the field. He made this part of the country's antiquities laws on archaeological and fossil sites, and it provides great experience for talented students to work alongside someone like geologist Frank Brown, or Rick Potts, who did the taphonomic research at Olorgesailie. As mentioned earlier, this educational program was part of Richard's plan for developing the small Nairobi Museum where he began to work in 1968.

Twenty-one years later, in 1989, Richard resigned as director of the National Museum of Kenya at the request of government officials. Richard, as head of the East African Wildlife Society, had voiced his concern about government employees who were involved in poaching elephants for their ivory tusks.

FROM FOSSILS TO TUSKS

At the time, the slaughter was rampant. In only four months poachers had killed at least 90 elephants in Kenya. A decade ago there were 65,000 elephants; in 1989 there were only 18,000. At that rate, there would be no elephants left anywhere in the country by the year 2000. The poachers were equipped with automatic rifles; they shot several tourists as well as the much-loved lion conservationist George Adamson, who got in their way.

Moving ivory across the country to the coast for transport north by dhow boats requires vehicles and a network, much like the illicit drug trade of South American countries. In the past, that network has included "game wardens" who turned a blind eye for a piece of the profit, as well as vehicles owned by a game depart-

Richard Leakey with his Cessna 206, which he uses to fly to remote fossil sites in Kenya. Today Leakey serves as head of the Kenya Wildlife Service and has led a successful war against elephant poachers. (© Delta Willis)

ment official. A list of those involved had been submitted to the Kenya government and filed. Leakey did not submit the list, but he openly demanded that the government do something about it. In September of 1988, he told the press that government officials were involved in the poaching. He knew that media exposure would make the government officials take action; their first reaction was to ask for his resignation. The international outrage over Leakey's resignation sent the matter all the way to the top, to the Kenya State House.

In April of 1989, Kenya's President Daniel arap Moi appointed Leakey as the new head of the country's game department, in charge of all the national parks and, more urgently, of stopping the poachers. The murder of an American woman on safari in Kenya had made headlines in the United States Moi's chief interest was to avoid bad publicity that discourages visitors. Tourism is Kenya's number-one source of foreign currency. Leakey and other conservationists argued that if the elephant disappeared, the tourist trade would suffer. In 1988, 700,000 foreigners went on safari in Kenya, bringing $400 million into the country.

If the Leakeys relish challenges, Richard had one of the most difficult and dangerous challenges in his life—which he found "exciting." In January of 1990 the *New York Times* Sunday Magazine ran a story filed by their correspondent in Nairobi: "Can He Save the Elephants?" Less than a year later, Richard Leakey had made remarkable strides toward saving Kenya's elephant's from poachers.

But like his leadership role in fossil discoveries, many people were involved behind the scenes. A group of ecologists and conservationists lobbied for an international ban on ivory trade. This made the price of ivory drop and the poachers' work not worth the risk. President Moi decreed that poachers could be shot on sight. Many were killed, but even more were held for questioning, to find out who was behind their network. Leakey fired hundreds of employees in the Kenya game department, who were corrupt on various counts besides being implicated in the poaching. As a result, he and his family received several death threats, and Leakey is today accompanied by armed bodyguards. He locks up his airplane inside a hanger rather than leaving it outside, and he rides in one of two vehicles, with one as a decoy. "It's a wonder he hasn't been killed," says anthropologist Daniel Stiles. "Certainly the threat of death was why no one else was brave enough to fire those people." The threats were reminiscent of Louis and Mary's experience during the "Mau Mau" rebellion.

Meave prefers to work in the field rather than be at their home or around Nairobi; she continues to lead the fossil research at West Turkana today.

Richard joins her when he can, but for the first year in his new job, his schedule was even more hectic than before. He found himself working very hard on five hours of sleep a night. In addition to reorganizing the parks administration and anti-poaching efforts, he appeared frequently on U.S. talks shows, such as "Good Morning America," appealing to viewers not to buy ivory jewelry. Until then, the United States imported $20 million of raw and carved ivory annually. He also made some rather bold statements about how conservation should work. The animals should pay for themselves; receipts from park entrance fees should go back into conservation, and fences may be required around park boundaries, to diminish the conflicts between wild animals and the fastest-growing human population in the world.

Just as Louis had introduced Kikuyu to the curriculum at Cambridge, Richard found creative ways to solve problems. "Everyone was saying I had to stop the poaching, but I couldn't stop the poaching unless I could get the men into the field." The game department vehicles were run down and virtually useless. So Leakey talked a Nairobi car dealer into giving him 18 Land Rovers on credit. The collateral for the vehicles was a storehouse of ivory tusks that had been confiscated from poachers. A few months later the 12 tons of ivory were burned in a public protest against the ivory trade; President Moi put the torch to the pile. Leakey explains, "If I'd waited until I had the money to buy the Land Rovers, there might not have been any elephants left to save."

"What separates us from other animals," Leakey told a reporter from *U.S. News & World Report*, "is the ability to think of tomorrow in terms of yesterday." Leakey's view of our past gives him a edge to his work as director of the newly created Kenya Wildlife Service. "I satisfied my ego a long time ago with fossils," he said.

"I think the knowledge exists to conserve everything," he had said several years earlier. "We can conserve the soil and the forests and the water. We know how to do it, and we know why we should do it. Conservation means management, and the management of our own species is as important as management of elephant and rhino."

Extinct species are common in the fossil record. Those extinctions were caused naturally, not as a result of a single species dominating the earth. It is both ironic and fitting that Richard Leakey, who helped uncover the path of human evolution, now has to contend with managing the many descendants of his fossil discoveries.

Richard Leakey continues to contribute his ideas on the subject of human origins as well as conservation efforts, lecturing to U.S. audiences at museums and universities. His tours also include fund-raising efforts for the Kenya Wildlife Service. Like most things he plans, the budget is an ambitious $200 million.

In September of 1992, Doubleday published his fifth book, *Origins Reconsidered*, a sequel to the best-seller *Origins*, co-authored with science writer Roger Lewin:

From *Origins*:

> *Humans are the first animals capable of manipulating the global environment to a substantial degree. . . . There is now a critical need for a deep awareness that, no matter how special we are as an animal, we are still part of the greater balance of nature. Unless we achieve such awareness, the answer to the question of when the human species might disappear will be: sooner rather than later.*

CHAPTER 10 NOTES

p. 118 "As a family . . ." Richard Leakey, *One Life*, p. 21.

p. 118 "Even bitter rivals find room . . ." John Noble Wilford, "The Leakeys, A Towering Reputation," *New York Times*, October 30, 1984.

p. 118 "first family" *Science Digest*, August, 1984, p. 256.

p. 120 "would go off for no good reason . . ." Mary Leakey on partnership with Louis, as quoted in Richard Milner, *The Encyclopedia of Evolution*, p. 269.

p. 120 "I thought he was a very vital . . ." Mary Leakey, interview with the author, January 4, 1983.

p. 120 "I'll leave this to you," Richard Leakey, interview with author, October 14, 1982.

p. 120 "It's not the credit I want . . ." Meave Leakey as quoted in Delta Willis, *The Hominid Gang*, pp. 144–45.

p. 121 "We were a bit tense . . ." Richard Leakey, interview with the author, October 14, 1982.

p. 124 "exciting", Richard Leakey as quoted in William F. Allman, "Endangered Species," *U.S. News & World Report*, October 2, 1989, p. 61.

p. 124 "It's a wonder . . ." Daniel Stiles, interview with the author, December 2, 1990.

p. 125 "If I'd waited until I had the money . . ." Richard Leakey as quoted in Eric Ransdell, "The Leakey Offensive," *Outside*, January 1990, p. 40.

p. 125 "What separates us from the other animals . . ." Richard Leakey as quoted in *U.S. News & World Report*, October 2, 1989, p. 61.

p. 125 "I think the knowledge exists . . ." Richard Leakey as quoted in Delta Willis, *The Hominid Gang*, pp. 323–24.

p. 126 "Humans are the first animals capable . . ." Richard Leakey and Roger Lewin, *Origins*, p. 256.

GLOSSARY

bed: a layer of sediments, distinguished from layers above and below by time and mineral composition.

cast: a copy, or mold, of a fossil, made with fiberglass and latex.

deposit: any earthly material that accumulates, including sediments, minerals, soil, and volcanic and plant products.

fault: a crack in the earth's crust where there has been movement.

geology: the study of the earth's changes and composition. Paleontology and stratigraphy are geological studies.

hominid: humanlike primates that walk (or walked) upright, members of the family *Hominidae*.

hominoid: a term including hominids and apes that move (or moved) on four legs. Hominids are included in the family of hominoids.

lava: the molten material that flows during a volcanic eruption and later forms into a rock such as basalt, obsidian or pumice.

matrix: the soil and mineral rocks surrounding a fossil. The term literally means "womb."

paleoanthropology: the study of human ancestors.

paleontology: the study of plant and animals that lived in the past.

primate: the order of mammals that includes humans, apes, monkeys, lemurs, pottos, bushbabies, lorises and tree shrews.

primatology: the study of primates.

sediments: deposits of soil, minerals and eroded rocks; sediments can be deposited by a river, or moved by winds during erosion and deposition.

stratigraphy: the study of rocks, sediments and volcanic products, especially the order of the layers, or beds, they form. The stratigraphy of a landscape can be revealed naturally by erosion and fault movement or artificially by digging trenches.

tuff: volcanic ash that accumulates in layers.

washed: when buried fossils emerge on the surface of the earth as a natural result of eroding winds or rains.

FURTHER READING

Allman, William F. "Endangered Species," *U.S. News & World Report*, October 2, 1989, pp. 52–61.

Darwin, Charles. *The Descent of Man*. London: John Murray, 1871.

———. *On the Origin of the Species by Natural Selection*. London: John Murray, 1860.

Johanson, Donald, and Maitland Edey. *Lucy*. New York: Simon & Schuster, 1981.

Johanson, Donald, and James Shreeve. *Lucy's Child*. New York: William Morrow, 1989.

Lambert, David, and the Diagram Group. *The Field Guide to Early Man*. New York: Facts On File, 1987.

Leakey, Louis S. B. *Animals of East Africa*. Washington, D.C.: National Geographic Society, 1969.

———. "Family in Search of Man," *National Geographic*, 143, no. 1 (January 1973): 143–144.

———. "Finding the World's Earliest Man," *National Geographic*, 118, no. 3 (September 1960): 421–435.

———. *White African*. Cambridge, Mass.: Schenkman Books, 1966.

Leakey, Mary. *Africa's Vanishing Art*. New York: Doubleday, 1983.

———. *Disclosing the Past*. New York: Doubleday, 1984.

———. "Preserving Africa's Vanishing Art," *Science Digest* (August 1984): 56–63.

Leakey, Richard. *One Life*, Salem, N.H.: Salem House, 1984.

Leakey, Richard, and Roger Lewin. *Origins*. London: Macdonald & Jane's, 1977.

Leakey, Richard, and Alan Walker. "*Homo erectus* Unearthed," *National Geographic*, 168, no. 5 (November 1985): 625–629.

Lewin, Roger. *Bones of Contention*. New York: Simon & Schuster, 1987.

Lyell, Charles. *Principles of Geology*, 3 vols. London: John Murray, 1830.

Milner, Richard. *The Encyclopedia of Evolution.* New York: Facts On File, 1991.

Payne, Melvin. "The Leakey Tradition Lives On," *National Geographic,* 143, no. 1 (January 1973): 143–144.

Ransdell, Eric. "The Leakey Offensive," *Outside* (January 1990): 37–41.

Reader, John. *Missing Links.* New York: Penguin, 1989.

Weaver, Kenneth F. "The Search for Our Ancestors," *National Geographic,* 168, no. 5 (November 1985): 561–623.

Wilford, John Noble. "The Leakeys: A Towering Reputation," *New York Times,* October 30, 1984.

———. "New Skull Finding Challenges View," *New York Times,* August 7, 1986.

Willis, Delta. *The Hominid Gang,* New York: Penguin, 1991.

INDEX